STRATEGIC ISSUE PAPERS

SIPRI's new series *Strategic Issue Papers* focuses on topical issues of significance for the future of international peace and security. *The ABM Treaty: To Defend or Not to Defend?* is the first in the series, coinciding with the third ABM Treaty Review Conference. Further studies will address problems relating to arms reduction, the spread of arms, military and political strategy and the impact of technology on the conduct of peaceful East-West relations. The books will be concise, with short production times so as to make a timely input into current debates. They will be available simultaneously in paperback and hardback volumes.

sipri
Stockholm International Peace Research Institute

Sipri is an independent institute for research into problems of peace and conflict, especially those of disarmament and arms regulation. It was established in 1966 to commemorate Sweden's 150 years of unbroken peace.

The Institute is mainly financed by the Swedish Parliament. The staff, the Governing Board and the Scientific Council are international.

The Governing Board and the Scientific Council are not responsible for the views expressed in the publications of the Institute

Governing Board
Ambassador Ernst Michanek, Chairman (Sweden)
Egon Bahr (Federal Republic of Germany)
Professor Francesco Calogero (Italy)
Dr Max Jakobson (Finland)
Professor Dr Karlheinz Lohs (German Democratic Republic)
Professor Emma Rothschild (United Kingdom)
Sir Brian Urquhart (United Kingdom)
The Director

Director
Dr Walther Stützle (Federal Republic of Germany)

sipri
Stockholm International Peace Research Institute

Pipers Väg 28, S-171 73 Solna, Sweden
Cable: PEACERESEARCH
Telephone: 46 8/55 97 00

The ABM Treaty: To Defend or Not to Defend?

Edited by
Walther Stützle, Bhupendra Jasani
and Regina Cowen

Stockholm International Peace Research Institute

Oxford . New York
OXFORD UNIVERSITY PRESS
1987

Oxford University Press, Walton Street, Oxford OX2 6DP
Oxford New York Toronto
Delhi Bombay Calcutta Madras Karachi
Petaling Jaya Singapore Hong Kong Tokyo
Nairobi Dar es Salaam Cape Town
Melbourne Auckland
and associated companies in
Beirut Berlin Ibadan Nicosia

Oxford is a trademark of Oxford University Press

Published in the United States
by Oxford University Press, New York

British Library Cataloguing in Publication Data
The ABM treaty: to defend or not to defend?
 —(Strategic issue papers).
  1. Arms control
  I. Stützle, Walther   II. Bhupendra, Jasani
  III. Cowen, Regina   IV. Stockholm
  International Peace Research Institute
  V. Series
327.1'74   JX1974
ISBN 0-19-829123-X
ISBN 0-19-829119-1 Pbk

Printed and bound in
Great Britain by Biddles Ltd,
Guildford and King's Lynn

Preface

The foremost challenge of this age and the next is to harness our political resources to prevent the East-West relationship from turning into a conflict. Among the measures employed to stabilize the strategic relationship, the 1972 Anti-Ballistic Missile (ABM) Treaty is of profound importance. It codifies the central principle of the nuclear age: a co-operative approach in limiting offensive *and* defensive strategic forces.

This principle has been challenged—both by failures in East and West to resist the allure of technology and by a more general erosion of political will to tackle together the problems of the nuclear offence-defence relationship.

This study makes a contribution to the debate over the strategic relationship, in particular the role of the ABM Treaty.

It is the first in a series of *Strategic Issue Papers* (SIP) that address topical issues of international peace and security. In order to contribute to a better understanding of the problems involved and to stimulate debate, SIPRI has produced this book before the commencement of the third ABM Treaty Review Conference.

The SIPRI editors would like to take this opportunity to thank Gillian Stanbridge for her outstanding editorial assistance. Furthermore, to ensure timely publication, she and Åsa Pihlstrand produced the camera-ready pages at SIPRI.

SIPRI　　　　　　　　　　　　　　**Dr Walther Stützle**
June 1987　　　　　　　　　　　　　　**Director**

Contents

Paper 3. The Treaty's basic provisions: view of the Soviet negotiator
Vladimir Semenov

Part III. Current US and Soviet ballistic missile defence programmes

Paper 4. Soviet research and development of directed-energy weapons
Simon Kassell

Paper 5. Technologies of ballistic missile defence
Thomas H. Johnson

Part IV. ABM Treaty issues—US and Soviet views on the Treaty and compliance questions

Paper 6. Issues related to current US and Soviet views of the Treaty: a Soviet jurist's perspective
Vladlen S. Vereshchetin

Paper 7. The ABM Treaty: verification and compliance issues
Richard N. Haass

Part V. The international dimension

Paper 8. Ballistic missile defences into the next century—implications for the United Kingdom's strategic deterrent forces
Ronald Mason

Paper 9. The implications for France
Dominique Moïsi

Part I. Introduction

The challenge to the ABM Treaty

Walther Stützle, Bhupendra Jasani and Regina Cowen
Stockholm International Peace Research Institute

I. Introduction

The challenges to the 1972 ABM Treaty are many and complex. Under threat is the original and continued validity of the strategic and political assumptions that provided the bed-rock for the Treaty back in 1972. The basis for all such assumptions was that mutual vulnerability held the key to mutual survival in the nuclear age; the immensely destructive power of nuclear weapons is a grim fact of strategic life. From this followed the recognition by the United States, and eventually shared by the Soviet Union, that nation-wide defences against ballistic missiles would most probably cause a destabilizing offence/defence arms race undermining the powers' individual and joint ability to control a crisis situation. The ABM Treaty was signed and ratified because it reflected the political determination of both sides to grapple with the realities of nuclear weapons and the requirements of strategic stability.

As a treaty, the ABM agreement is not perfect. For example, it does not contain a definition of 'development' and 'sub-component', nor does it cover anti-satellite (ASAT) and anti-tactical-ballistic missile (ATBM) technology. Moreover, since 1972, technology has not stood still, neither in the offensive nor in the defensive field. This necessitates clarification of treaty provisions in order to close emerging technological loopholes. This is not surprising—few treaties can lay claim to indefinite perfection. The negotiators of the ABM Treaty foresaw the need for possible amendments to treaty regulations and provided for it. What is surprising is the apparent inability of the United States and the Soviet Union to resolve problems of definition, interpretation and compliance in such a way that the ABM Treaty and the strategic rationale it represents can survive.

The strategic relationship is at a crossroads. Sustained political commitment by the US and the USSR and their respective allies is needed in order to reaffirm the principle of mutual security—the promise of strategic defences raises false hopes.

In the autumn of 1987 Government representatives from the United States and the Soviet Union gather for the third Review Conference of the ABM Treaty. Unlike previous review conferences, the next one could well become a milestone in the history of the Treaty or speed the process of its continued erosion and eventual demise. Yet the Conference itself cannot provide the only forum to resolve the legal, technological and political challenges to the ABM Treaty; the issues are already too contentious and politically polarized. Nor can the Conference, without political will by both sides, restore mutual confidence in the basic accord that led to the conclusion of the Treaty 15 years ago. What the Conference can do, is facilitate the emergence of a sensibly conducted political discussion on the relevance of strategic defences and their bearing on the ABM Treaty. The prospects for this to happen, however, are not good.

This introductory chapter will attempt to shed some light upon the network of single and multiple challenges to the ABM Treaty in order to provide insight into an on-going political dispute. The resolution of this dispute will in one way or another undoubtedly be of paramount importance for the future of East-West relations and international peace and security.

II. The strategic relationship

Article XV, paragraph 1, calls the Treaty one of 'unlimited duration'.[1] Paragraph 2 allows each party 'the right to withdraw from this Treaty if it decides that extraordinary events related to the subject matter of this Treaty have jeopardized its supreme interests'. Have such 'extraordinary events' warranting a US withdrawal from the Treaty occurred? Or has the strategic rationale that underpins the Treaty somehow lost its validity—a rationale that assumes mutual assured survival to rest upon deterrence based on offensive nuclear forces? The answer to both questions is no—despite US Administration pronouncements to this effect and continued Soviet efforts in defensive technolgies.

The inescapable fact is that both sides have the capacity to inflict unacceptable levels of damage upon each other, despite improvements in defensive technology. This has brought about a relationship of mutual dependence unknown in the history of states. In an age of strategic parity, independent national security planning has lost much of its meaning. Security must be defined bilaterally; it is no longer a question of unilateral concern. Unilateral attempts to strengthen deterrence through some form of ballistic missile defence (BMD) deployment make little sense. The concept of deterrence itself is derived from the superiority of the offence. If this superiority is challenged or even undermined or eroded, deterrence itself will lose credibility, taking with it all the military and political strategies based upon it. Deterrence through retaliation is not an easy concept to live with; it is anathema to many people's personal values to threaten another country's population with annihilation in order to protect their own well-being. Yet a security system based upon assured retaliation is inherently more stable than one based on a mix of offensive and defensive forces. This requires some elaboration.

Strategic stability is presently understood to exist when neither side has an incentive to launch a first strike against the other. It also means that one side could absorb a first strike and still have the ability to retaliate causing unacceptable levels of damage to the attacker. In 1983, the Scowcroft Commission defined strategic stability in these terms.[2] If defensive systems were to be introduced by one side, that is unilaterally, strategic stability would be in jeopardy. In a crisis situation, unilaterally deployed defences provide an incentive for both protagonists to strike first. The side without defences will want to use its previously designated retaliatory forces quickly to exploit the moment of surprise, to be sure that its forces are not decimated by a first strike by the other side. Only a first strike will ensure that its surviving retaliatory forces are not subsequently intercepted by the opponent's defences. The side which has deployed defences will want to use them to maximum effect. Familiar with the above scenario, it will launch a first strike with the intent of depriving the opponent of the benefit of surprise. The intention would be to disable a substantial part of the adversary's forces, before they are launched, and to intercept the remainder. These examples are illustrative only and assume that even limited defences are highly capable. The point, however, is that unilaterally deployed defences have a bilaterally destabilizing effect: both sides will be tempted to pre-empt.

Even if both sides were to deploy strategic defences co-operatively, as envisaged by President Reagan, the destabilizing effects would remain. Each protagonist would have to assume that his defences were less capable than those of the other—every prudent military planner has to contemplate worst-case scenarios. Each side would want to ensure that its offensive forces maintained the ability to penetrate defences on the other side and would therefore want to increase its offensive capability. Most important, however, in a crisis situation both sides might be tempted to believe that there was a military advantage in striking first. Both sides would try to destroy the bulk of each other's retaliatory forces and then hope that their defences were able to intercept the opponent's remaining missiles. In fact, there is little difference between the destabilizing effects of unilaterally and co-operatively deployed defences. In both situations, the urge to pre-empt could be paramount.

Despite the overwhelming importance of crisis planning a military planner must also address issues of peace-time force planning. The introduction of defensive systems radically alters the peace-time confidence in the capability of one's forces. It could be argued that uncertainty about force effectiveness contributes positively to deterrence and stability. However, the real world has a habit of spoiling good theories. It is far more likely that instead of maintaining forces of questionable effectiveness, a military planner would compensate for this doubt by increasing military capabilities. It may be necessary to advise political leaders to deploy defensive systems both to offset the other side's technological challenge and to match its range of capabilities. As a result, both sides will race each other with offensive and defensive nuclear forces and deploy a combination of both.

The prospect of a future where both the Soviet Union and the United States take such steps, probably in order to protect their retaliatory forces and strengthen deterrence, should delude none. Why then is a treaty which successive US administrations since 1972 believed to be in the national security interest being challenged?

III. The Reagan challenge

By its actions and pronouncements the US Government has suggested that it no longer attaches much value to the ABM Treaty. In October 1985, the United States announced a new interpretation of the Treaty that permits development and testing of ABM systems 'based on other physical principles'.[3] The Administration has since argued that the new interpretation is legally correct but that for the time being, US policy would uphold the traditional 'narrow' Treaty interpretation.[4] The 1985 'broad' interpretation is mainly concerned with existing Treaty provisions on new, that is, post-1972, technology. Traditionally, the Treaty has been interpreted to allow all kinds of research, irrespective of the nature of the technology involved, while banning development, testing and deployment of systems and components that are not fixed, land-based (Article V, 1), and imposing strict limits on land-based ABM systems and components—Article III (a) and (b). The new interpretation takes issue with the Treaty's definition of the terms 'ABM systems and components'. Article II (1) states that an ABM system 'currently' consists of ABM interceptor missiles, ABM launchers and ABM radars. The new interpretation alleges that this definition of an ABM system in Article II (1) is specific rather than functional. In other words, even if a system of component is intended for an ABM role, development is allowed if the system or component is not specifically mentioned in the Treaty. Proponents of the new interpretation advance the view that Article II, 1 defines and applies only to systems and components based on then current technology but not to systems and components based on future technology. This reading of Article II (1) poses a serious threat to the Treaty as a whole. Wherever reference is made in the Treaty to 'ABM systems and their components', 1972 technology is banned or restricted and anything post-1972 would be permitted subject only to the deployment consultation clause in agreed statement D.

The legal case for the new treaty interpretation hinges upon agreed statement D, which makes the only direct Treaty reference to 'ABM systems based on other physical principles'. If, so the argument goes, Article II (1) were a functional definition applying to any system that can 'counter strategic missiles or their elements in flight trajectory', there would not have been any need for agreed statement D; both current and future technologies would have been dealt with already. Agreed statements serve the purpose of reinforcing the basic provisions of a Treaty. This is precisely what agreed statement D does. Its opening phrase, 'In order to insure fulfilment of the obligation not to deploy ABM systems and their components except as provided for in Article III of the Treaty...' reiterates the purpose of the Treaty. Moreover, agreed statement D lays down procedures for dealing with 'ABM systems based on other physical principles and including components capable of substituting for ABM interceptor missiles, ABM launchers, or ABM radars'. It states that 'specific limitations on such systems and their components would be subject to discussion in accordance with Article XIII and agreement in accordance with Article XIV of the Treaty' (see the contributions by the Treaty negotiators Smith and Semenov, papers 2 and 3). The US reinterpretation was not preceded by 'discussion' with the Soviet Union; this omission does not facilitate subsequent agreement on 'specific limitations'.

The new interpretation not only challenges specific provisions in the Treaty but undermines its basic purpose, which was to close off the path to an arms race with defensive weapons. The new reading makes a mockery of the ban on development, testing and deployment of ABM systems or components which are sea-based, air-based or mobile land-based listed in Article V. It lifts the prohibition on transfer of ABM systems and components laid down in Article IX to the extent that they are based on 'new physical principles' and would permit the United States to involve its allies in outright development and testing of strategic defensive systems. It is difficult to imagine that such revisionism will not turn the Treaty on its head.

Judge Abraham Sofaer, legal adviser to the State Department and primary architect of the new interpretation, has correctly called attention to existing Treaty ambiguities.[5] Yet, once such ambiguities have been identified, they should not be of immediate political relevance. If they were, it could be plausibly argued that the Nixon, Ford and Carter administrations had failed dramatically in providing for the security of the nation. With the conclusion of the ABM Treaty, the US Administration should have initiated a rigorous BMD research programme into systems based on new physical principles. The Reagan Administration is implicitly charging previous administrations with gross negligence if not outright incompetence. Rather than using legal ambiguities as a rationale for a radical revision of the Treaty, the Reagan Administration could have used existing channels of communication for discussing legal uncertainties with the Soviets. The Standing Consultative Commission, established through Article XIII, would have been the most obvious body for clarification and amendment of Treaty provisions. But the Reagan Administration's arms control policy and its view of the strategic relationship do not dictate such steps.

It has almost become a cliché that Ronald Reagan came to the White House after successfully campaigning against a SALT II treaty he called 'fatally flawed'. Cliché or not, Ronald Reagan's view on the future course of US arms control policy epitomized the belief among many that a restoration of pride in US values had to be translated into strong and steadfast US leadership—the post-Vietnam era had not just come to an end, with Ronald Reagan it had become a glorious beginning. Not everyone shared in the revival of US leadership with equal enthusiasm.

US allies in Europe who had deplored the indecisiveness of the Carter Administration soon realized that President Reagan's political approach to East-West relations was at least as much based on deeply held views of the Soviet Union and its political system as it was guided by the necessities of superpower responsibility. During the first Reagan Administration the political imagery began to undergo dramatic change: the East-West détente of the 1970s had not prevented Russian and Cuban involvement in civil strife in Africa; the 1979 Soviet invasion of Afghanistan was portrayed as the latest and final proof that détente had failed; arms control had not put any reins on the Soviet military buildup and was therefore no longer an appropriate policy tool for the United States in its relations with the Soviet Union. It was only heavy pressure from the public at home and allies abroad that persuaded the President to once again discuss arms reductions with the Soviets. Again, arms control appeared to be a useful political tool, even if only to placate criticism within one's own camp.

1983 was the year that President Reagan launched his Strategic Defense Initiative (SDI). While calling upon the progressive forces of technology and innovation, SDI is very much an exercise in political and strategic nostalgia. It is *political* nostalgia because it conjures up a modern version of military superiority, using the advances in technology to achieve what can no longer be gained through policy. It recalls the early post-war years when Americans could be magnanimous without suffering the burdens of great power responsibility and superpower competition. The President's offer to share SDI technology with the Soviets is an additional example of this political time-warp. It is *strategic* nostalgia because it dismisses a military reality which does not give many options to policy-makers. Mutual assured destruction, as discussed above, is a feature of the strategic relationship, not a policy choice. In his SDI speech, the President entertained the possibility that there might be other options than MAD. Perhaps there are. The point about the existing strategic reality which was noticeably absent from the President's speech is the inherent implications for national security planning imposed by the immensely destructive power of nuclear weapons. The Reagan Administration, in its efforts to restore US military superiority and political confidence has neglected the basic premises of the nuclear age that prescribe the search for a co-operative relationship with the Soviet Union.

To understand this neglect, one should remember that the 'Reagan revolution' reflects and builds upon an upsurge of American nationalism. The extent to which relations between the United States and the Soviet Union have been affected by this is well known. The new Reagan Administration viewed arms control as a sign or admittance of weakness, an excuse for not standing up for the things one believes in, an indirect recognition of the Soviets as equal partners, and an endorsement of Soviet strength instead of American virtue. Terms such as compromise and negotiations were anathema to the perceived need for US self-assertion. Much of this thinking has been institutionalized in the Reagan presidency and is unlikely to change during its remaining years.

As the substance of US security thinking has swung towards unilateralism and away from the traditional understanding of deterrence, so has the style of the Reagan Administration. Reagan's own political style is individualistic and spontaneous, relying on dramatic intervention in order to break an apparent deadlock. His SDI speech is as much an example of this as is his policy towards Iran. SDI challenges the perceived deadlock of mutual assured destruction, just as the sale of arms to Iran in exchange for hostages was meant to break the deadlock in dealing with international terrorism. At the Reykjavik Summit, President Reagan offered to eliminate ballistic missiles, showing a single-handed readiness to abandon a defence policy adopted by 16 Nato countries. Since 1983, the President has persisted in not offering SDI as a bargaining tool to negotiators at Geneva. It may well be the case that SDI persuaded the Soviets to return to Geneva but that event alone does not mean an arms control agreement is in the offing. The President has thus far not been able to persuade the Soviets to make cuts in offensive strategic forces. US allies have also had cause for concern. While they have been accustomed to a certain degree of unilateralism in US dealings with the Soviet Union, they are far from ready to view Western defence policy formulation as a unilateral US prerogative. The Strategic Defense Initiative was not discussed with the allies prior to the President's speech, nor were they privy to the President's stance at Reykjavik. Most of all, however, US allies feel that the US approach to the Soviet Union lacks coherence.

In May 1986, the President announced that the United States Government would no longer abide by the unratified SALT II treaty.[6] This decision seems to have been based more on a short-term reaction to Soviet non-compliance than on a compelling analysis of the national security implications of the Soviet violations *and* the cost to the United States if there were no constraints on the Soviet military buildup. At the end of November 1986, the US deployed the 131st B-52 armed with air-launched cruise missiles (ALCMs), abrogating the previous US commitment to abide by the provisions of the unratified SALT II agreement.[7] With SALT II no longer in operation, the technical demands made upon the Strategic Defense Initiative have immediately become less predictable and therefore far greater than would have been the case if SALT-type constraints on Soviet offensive forces had continued to provide a basis for calculation. On 20 February 1985, White House arms control adviser Paul Nitze, in a speech to the World Affairs Council of Philadelphia, laid down criteria for deployment of SDI systems: systems survivability and cost-effectiveness 'at the margin'.[8] In the absence of an agreement on offensive force limitations, it is difficult to see how Nitze's deployment conditions can be met. In order to be effective, defensive systems must interact successfully with offensive systems. They must be survivable as a whole, not merely in part; they must be cheap enough to be improved at reasonable cost so as not to be rendered impotent through an incremental increase in offensive capability; and, they must not lead to a destabilizing offence/defence arms race. In any case, however, the Nitze guide-lines no longer appear to be decisive in official US BMD deployment thinking. Neither do repeated assurances to US allies that the SDI programme will be structured in accordance with US obligations under the ABM Treaty. Towards the end of 1986, Administration officials, notably Defense Secretary Caspar Weinberger, began a concerted push for 'phased deployment' of BMD systems in the early 1990s.[9] The reasons for speedy deployment are mostly political.

1. SDI supporters perceive a myriad of challenges to continued large annual SDI budget increases. Congress is unlikely to fund a mere research effort indefinitely; positive research findings need to be shown to exist in order to maintain the belief in the technical feasibility of SDI.

2. The Reagan presidency, weakened by the Iran affair, is running out of politically useful time. If ever, the stage for BMD deployment must be set now in the hope of making a positive deployment decision irreversible. While the next president may not be as convinced of the merit of BMD as Ronald Reagan and opt for a sharply reduced SDI budget and an extended research time frame, if the ABM Treaty can be broken decisively within the next few months, an irreversible event will indeed have taken place.

The technologies being considered for early deployment, that is during the first phase, are by and large not based on 'new physical principles'. The US Department of Defense has gone to great lengths to demonstrate the feasibility of rapid deployment. In September 1986 a Delta rocket was launched carrying two test vehicles; one was the second stage of the Delta, the other an SDI target satellite. The second stage was put in orbit and the target satellite separated thereafter. The Delta sensors observed the characteristics and behaviour of the satellite in space. Subsequently, a Minuteman missile second stage was launched and its infra-red exhaust plume was observed by the orbiting Delta launcher. The second stage of the Delta then tracked the satellite and destroyed it in a head-on collision.[10] Although an important experiment, it hardly qualifies as a technological breakthrough of a magnitude that would allow a deployment decision. Kinetic energy devices, which kill through impact, even if equipped with sophisticated sensors, do not solve the problem of mid-course discrimination between warheads and decoys and are themselves highly vulnerable to attack, if space-based. Without an anti-satellite (ASAT) agreement, any object in orbit is a sitting duck—an ASAT agreement is, however, as unlikely to be concluded as a new ABM Treaty should the present one cease to exist. Indeed, in light of the extensive similarities between BMD and ASAT technologies, an ASAT agreement appears positively undesirable to those favouring an early deployment decision. The technologies of ASAT and BMD are very similar; but since it is relatively easier to disable a satellite than to destroy an ICBM missile or warhead, the physics of the kill mechanism required makes greater demands in the case of ballistic missile defences. Some SDI-related experiments can thus be conducted under an ASAT guise without apparent violation of the ABM Treaty. While such fanciful disguises are possible, there should be no question as to their combined effects: they erode the basis and substance of the ABM Treaty and speed the process of its demise.

On the whole, the Delta experiment cannot escape a certain irony. The Reagan Administration reinterpreted the ABM Treaty in order to facilitate SDI but is now considering initial deployment of non-exotic kinetic kill vehicles.

While there has been technological progress in SDI research, there has also been mounting evidence of the complex nature and size of the task at hand. The technological obstacles to achieving even limited BMD capability in space are nowhere close to being solved. There is a long way between a technologically promising idea and an operationally competent system. Moreover, the very schedule-driven nature of the SDI programme itself may not be in the long-term interests of the United States. BMD technologies still require a long-term research commitment, not a premature decision based on what looks promising now. Most of all, BMD research needs a stable environment. It needs an arms control regime stabilizing offensive forces as much as predictability in the defensive field. The Reagan Administration has already done much to weaken the former and is about to do away with the latter.

IV. The Soviet challenge

Much of the US Administration's attitude to arms control and the US–Soviet relationship has been generated domestically, yet Soviet arms control behaviour has at best been ambiguous and at worst non-compliant with Treaty obligations. Even among critics of the US Strategic Defense Initiative there is a consensus on Soviet ABM efforts; before and since 1972 they have been steady (see paper 4). The Soviet Union, as allowed by the Treaty, maintains a ring of defences around Moscow. This is currently being upgraded from 64 Galosh interceptor missiles to 100 SH-04 and SH-08 nuclear-tipped missiles.[11] The Soviets are also improving their radar capability; Article VI(b) commits both sides 'not to deploy in the future radars for early warning of strategic ballistic missile attack except at locations along the periphery of its national territory and oriented outward'. Six new large phased-array radars are being built to enhance Soviet capability to detect a missile attack. One of these radars being built at Krasnoyarsk in Siberia and widely believed to be an early-warning radar is not where it should be. It is neither on the national periphery nor does it point outward. Soviet claims that this radar is only for space tracking will not be substantiated until it is switched on.[12] Doubts about Soviet intentions, however, are with us today (see paper 7).

The Soviet Union is also conducting extensive research into kinetic, laser and particle beam technology. The US Government believes that the Soviets are developing components for rapidly deployable mobile ABM systems. A related problem is posed by the SAM upgrade issue. Article VI(a) of the ABM Treaty prohibits an upgrading of air defences to ABM capability. The Soviet SA-12, however, is believed to have some ABM capability, possibly against shorter-range ballistic missiles and perhaps also against submarine-launched ballistic missiles.[13]

The questions raised about Soviet BMD activities stem from persistent uncertainty about Soviet motives. Clearly, the technological obstacles faced by SDI must also be obvious to the Soviet Government in its BMD research. One explanation of Soviet defence research could be that it is a hedge against a US breakout from the Treaty. Another explanation, favoured by the US Government, could be that 'the aggregate of current Soviet ABM and ABM-related activities suggests that the USSR may be preparing an ABM defence of its national territory...'[14]. Undoubtedly, improvement of the ABM system around Moscow will make that system more effective; although it is unlikely to withstand anything but a small ballistic missile attack. The new phased-array radars will increase Soviet early-warning, attack-assessment and target-tracking capability. Because the Soviets have maintained a sustained ABM programme, they can be expected to have open production lines and personnel trained to operate the Galosh interceptor systems. Thus, while it can be assumed that Soviet scientists encounter similar problems to those of their US colleagues with exotic BMD technologies, the Soviets may have an edge should they become convinced of the necessity for rapid deployment of traditional fixed or new mobile land-based ABM systems. This may not actually happen, but recent US steps towards an early ABM deployment decision could affect Soviet thinking. Neither side would benefit from such developments. Even if the Soviets could put vast numbers of ABM systems into place, the sum of the parts may tell little about the effectiveness of the whole. The same would apply to US ABM deployments. Yet there is the possibility that both sides, because of existing uncertainty regarding the eventual motives of each, could force each other into deploying what both know not to be effective. Both sides would break the ABM Treaty because each believed the other was about to do so.

Presently, both sides are uncertain about each other's BMD motives and ultimate intentions; both sides are engaged in activities that raise ABM Treaty compliance questions. The Soviet Government, although it professes to uphold the narrow interpretation, has not come forth with sufficient evidence to dispel fears in the West about the extent of the Soviet BMD research and the range of research programmes. This has led to questioning of Soviet motives and a reassessment of Soviet interests in negotiated arms control generally.

In his much noted study of Soviet defence thinking, David Holloway discusses two strands in the Soviet approach to the problems of the nuclear age that have long bedeviled Western analysis of Soviet strategy: war deterrence and the preparation for war: 'The party leaders stress the defensive nature of Soviet military doctrine, but Soviet military strategy gives primary importance to the offensive, and the combination of a defensive doctrine and an offensive strategy has seemed at best ambiguous, at worst threatening, to Western governments'.[15]

Apart from the difficulties in the West with seeing an accurate distinction between Soviet doctrine and strategy, the two concepts are clearly at odds with one another. Strategic parity and its codification in the SALT treaties would suggest Soviet acceptance of strategic deterrence and the military *status quo*. But during previous arms control negotiations the Soviets have refused to reduce their inventory of heavy ICBMs and ameliorate the threat they pose to US ICBMs. The arms control negotiations of the 1970s appear to have had little impact on the momentum of the Soviet military buildup and the structure of Soviet forces.

Current Soviet force developments suggest across-the-board modernization and replacement of strategic forces.[16] The new road-mobile SS-25 ICBM has been deployed and the SS-X-24, a rail-mobile ICBM, is being flight-tested. A successor system to the SS-18 ICBM is also anticipated. While the Soviets are moving towards deployment of mobile ICBMs, they are likely to continue with silo-based forces. Soviet bomber forces are also being modernized. A new version of the Bear bomber, the Bear H, equipped with AS-15 air-launched cruise missiles has been operational since 1984. The Blackjack, a new strategic bomber, may soon be deployed with Soviet forces. Currently, the Soviets have deployed some 9000 strategic warheads. Through replacement of older systems and some changes in force structure, they could stay within the quantitative sub-limits of SALT II and still increase the number of warheads to around 12 000 by 1990. Given the breadth and depth of Soviet modernization programmes, it is believed that in the absence of SALT-type constraints the Soviets could deploy between 16 000 and 21 000 warheads by the mid-1990s. Since force modernization does not happen overnight but requires a sustained resource commitment over a number of years, it appears that the Soviet Union is following a long-term strategic planning policy. Strategic arms control thus far may have tempered the pace of the Soviet military buildup but does not seem to have had an impact on Soviet long-term modernization programmes. While there is no evidence at present of Soviet intentions to expand their strategic forces much beyond SALT II constraints, the potential for such expansion exists. The Soviets are unlikely to forgo major force modernization in order to increase investment in other sectors of the economy. Moscow's arms control record shows a strong commitment to preserve and enhance Soviet military capability in the offensive and defensive fields alike. By the early 1990s, the Soviet Union will be able to field new and expanded forces in all major categories including some 2000 to 3000 subsonic cruise missiles against which an SDI-type defence has little to no capability.

Force modernization in the United States has encountered a number of political and technical obstacles. Deployment of MX missiles in Minuteman-III silos is only a temporary solution. Soviet ICBM accuracy has increased, and is seen to endanger US silo-based forces. Mobile MXs and Midgetman missiles are still in the planning stages. The B-IB, the future strategic bomber, is not performing to technical specifications. Although the Trident-II SLBMs will make up the most survivable leg of the strategic triad, if ICBM vulnerability cannot be resolved, the case for strategic defences to decrease vulnerability of ICBM forces will be enhanced. This even more so, if Soviet cruise missile capabilities put the US strategic bomber forces at risk

In order to restrain the pace of Soviet across-the-board modernization of offensive forces and decrease US ICBM vulnerability, a new strategic arms accord is needed. It will need to substantially reduce the number of accurate warheads on Soviet heavy ICBMs and should address the MIRV[17] potential of the SS-25. An agreement on the limitation of offensive forces is more likely to have a positive impact upon the strategic balance if concluded before a potential Soviet proliferation of warheads takes place in the early 1990s. It does not seem very likely, however, that a substantial agreement on offensive forces can be reached unless there is a concomitant agreement on the direction of ballistic missile defensive programmes.

V. Future arms control

The technical synergism between offence and defence is well established and was reflected in the 1972 ABM Treaty. Yet the failure to control emerging offensive technologies has led to a questioning of the continued viability of negotiated arms control. At Reykjavik, both President Reagan and General Secretary Gorbachev seemingly discarded the traditionally incremental approach to arms control. Not just deep cuts but the elimination of ballistic missiles was on the table. This willingness of both leaders to leave behind established rules of procedure is an indication of the crisis traditional arms control is undergoing. The apparent ease with which the arms control stakes are increased, as though arms control sincerity could be measured by the radical nature of the proposal, shows the malaise of arms control thinking on both sides of the divide. A world with fewer nuclear weapons would not *automatically* be a safer world. Deep cuts in strategic forces would have to be carefully weighed against their impact on force structure and, especially in the case of the United States, defence commitments to Western Europe. This will not be easy and time is running short.

First, both sides should reaffirm arms control's basic role which is to facilitate stable strategic relations. The agreement in 1972 not to deploy nation-wide defences has strengthened deterrence in several ways: *(a)* it has prevented an arms race in offensive and defensive deployments; *(b)* it has assured the effectiveness of retaliatory forces; and *(c)* it has denied both sides a militarily significant degree of superiority. Since 1972, however, offensive forces have grown both in quantity and quality to such an extent that the vulnerability of ICBMs has brought strategic defences, as a means to protect ICBMs, back on the agenda. Over the same period, research into defensive technologies has progressed, particularly in kinetic energy and directed energy. Yet progress in these fields does not alter the essential strategic equation which still decidedly favours offensive nuclear weapons. Even if an effective defence of ICBM silos could be achieved, it would be highly undesirable. Defences against ballistic missiles raise fears of an emerging first-strike capability with all the familiar threats to strategic and crisis stability. Instead the United States and the Soviet Union should adopt a 'dual-track' approach to arms control. The present situation shows that arms control cannot survive on either an offensive or defensive leg alone. In order to preserve strategic stability, the control of offensive and defensive capabilities must not only be pursued simultaneously but these efforts must complement one another. In other words, it is not enough to have a variety of forums dealing with different force capabilities; cuts in offensive forces cannot be achieved in a meaningful way if discussions on defensive forces do not take into account that the effectiveness of offensive forces demands defensive force limitations. Likewise, the effectiveness of limitations on defensive forces rests upon limitation of offensive forces.

Second, the United States must understand that strategic arms control is in the US and Western interest. The Reagan Administration's concern with questions of Soviet Treaty compliance is important but not important enough to abrogate SALT II and jeopardize the ABM Treaty; the Soviets have kept to the principal limitations of SALT I and SALT II. Questions of compliance are presently more of political than military significance—they reduce confidence and erode agreements. The Soviets, on the other hand, should not expect questions of compliance to be politically unimportant because of their low military significance. In an atmosphere of mistrust, one that has characterized most of the post-war US-Soviet relationship, a series of small violations have the potential of finding themselves on the centre stage of the strategic relationship. The Soviet Union, with its offensive military strategy and defensive military doctrine that contribute to inherently ambiguous interpretation, cannot afford to pursue ambiguous behaviour as a policy. If 'glasnost' is to provide openness and transparence at the Soviet domestic level, a measure of 'glasnost' in Soviet international affairs would not come amiss.

The problem with these suggestions is that they assume that a measure of positive political will exists on both sides. If this can not be assumed, it is doubtful whether ABM Treaty-related issues of technology, interpretation and compliance can be resolved.

Perhaps the single most important contribution of the Strategic Defense Initiative is the growing appreciation of emerging technologies that could have application for defensive purposes. Prior to President Reagan's 1983 speech, around one billion dollars annually was spent on BMD research in the United States.[18] Research focused primarily upon low-altitude nuclear-armed interceptors and non-nuclear exo-atmospheric interceptors and laser technologies. The Strategic Defense Initiative harnessed existing programmes, increased the variety of research activities and centralized the direction of BMD research. Because of Soviet MIRVing of ballistic missiles in the 1970s, the militarily most attractive defence concept to the United States is boost-phase defence, that is intercepting ballistic missiles before they shed their warheads. At that stage, the exhaust plume of a ballistic missile facilitates identification and tracking. Missiles not thus disabled will release warheads and decoys, multiplying the targets the defence needs to engage. Defensive systems for boost-phase and mid-course interception will have to be space-based or, at a minimum, some components will have to be put into orbit. Not surprisingly therefore, much of SDI research since 1983 has grappled with the technological challenges that space-based defences pose.

Apart from the 1985 US reinterpretation of the ABM Treaty, emerging technologies have begun to create 'grey areas' that diffuse the dividing lines between permitted and restricted military capability. Some technology has become multi-functional; this applies especially to sensor technology. Other technology is hovering precariously between an air-defence and an ABM capability. Moreover, components of other technologies, such as anti-satellite technology, can be used in BMD systems. Even without the broader ABM Treaty interpretation, these technical developments endanger the future of the Treaty. The new US interpretation speeded up this process politically, but technological advances would by themselves have forced a collision with the ABM Treaty. For example, large phased-array radars (LPARs) are crucial components of ABM systems. They provide early warning and battle management for ballistic missile defence. These LPARs are also used for space tracking. Article VI (a) of the ABM Treaty assumes that a functional distinction between an ABM radar and a non-ABM radar can be made with confidence. However, unless an LPAR also fulfils the location and orientation requirements of Article VI(b) it becomes very difficult to determine its precise purpose, since an LPAR could then be used in an ABM mode. Air-defence missiles such as the previously mentioned Soviet SA-112 but also US Patriot and Improved Hawk systems are believed to have a latent anti-tactical defence capability.

European defence initiative

There has been much debate within the NATO alliance about the stance European allies have adopted regarding SDI and the ATBM question (see papers 8, 9 and 11). Article IX of the ABM Treaty prohibits the transfer of ABM systems and components and their deployment outside the parties' territory. This Treaty provision takes care of outright transfers but does not prevent European participation in the SDI research programme. Despite the threats to the ABM Treaty posed by the unilateral US reinterpretation, the technological advances that increasingly blur the dividing lines between ABM and ATBM technologies and the already complicated compliance issues, West

European governments want a share of the SDI research pie and maintenance of the ABM Treaty. These two objectives are difficult to reconcile, to say the least.

The reasons for European SDI participation are political, economic and military. Support for SDI demonstrates alliance cohesiveness at a time when there is no real agreement in the alliance on how to respond to apparent changes in Soviet domestic and foreign policies. SDI is a major research initiative that is pushing back the frontiers of technology. West Europeans, already aware of the US technological edge, do not want to miss out on potential technology gains. The military rationale for participation in SDI is weak although it is most often used to justify European interests in ATBM technology. The principal argument is that NATO Europe needs an ATBM capability to deter the Soviet use of short-range ballistic missiles armed, at some point in the future, with conventional and chemical warheads.[19] It is entirely possible that the Soviets will produce non-nuclear warheads accurate enough to substitute for nuclear warheads. At issue in the ATBM debate should not, however, be Soviet technological capability but the precise military implication for NATO. There is no compelling need for NATO to respond to a Soviet conventional/chemical ballistic missile capability with an ATBM system. Since the main advantages in using non-nuclear ballistic missiles is their accuracy and penetration, they would most likely be used to destroy or disable key NATO facilities like missile sites, command and control centres and airfields. As a response to this threat, NATO should in the first instance investigate opportunities for reducing dispersal times, increasing command and control mobility and hardening aircraft shelters. This would greatly reduce the number of targets the Soviets could engage. It would also be a great deal cheaper than an ATBM defence, which has unknown deployment requirements: space-based and/or land-based? Finally, even if the Soviets acquire a non-nuclear capability, not all its tactical ballistic missiles may be thus equipped. A residual number of missiles would probably still carry nuclear warheads. How would NATO distinguish between different kinds of warheads? If NATO successfully intercepts a nuclear-tipped missile, the destruction of the warhead could free radioactive plutonium and contaminate allied territory, particularly if the Soviets decided to use impact fusing. Interception of chemical warheads would entail similar risks. However, whether Europeans would die quickly because of a direct nuclear hit or slowly through radioactive or chemical poisoning is not the most immediate issue in the ATBM debate. It is most urgent to realize that the whole question of defences against shorter-range systems has a direct bearing upon continual confidence in the ABM Treaty.

When talking about ATBM systems and participation in SDI, Europeans should be aware that they have certainly become party to the continuing erosion of the ABM Treaty. ATBM and ABM technologies display similarities that make technical distinctions difficult if not impossible. This ambiguity is exacerbated by deliberate political efforts to exploit the absence of ATBM limitations in the ABM Treaty. Although there are only two signatories to the Treaty, it would be against the intention and spirit of the Treaty for the European allies of the United States to undermine an agreement that has provided for their very security. It is highly questionable whether extended deterrence would be here today had the ABM Treaty not been signed in 1972. A United States that cannot effectively retaliate because of Soviet strategic defences is unlikely to uphold defence commitments abroad. If Europeans are truly committed to upholding the ABM Treaty; and their security situation makes it imperative that they are; they must recognize that quite a significant threat to the ABM Treaty stems from their own pursuit of contradictory objectives. A simple reiteration of the responsibilities of the two superpowers will not do; this may indeed be an issue that demands political coherence, diplomatic skill and an understanding of the facts on the part of the West Europeans.

Strengthening the ABM regime

One critical ABM Treaty-related problem area remains to be discussed. It epitomizes the political, legal and technological issues raised and will have to be a major item on the Review Conference agenda if the Treaty is to have any chance of surviving even over the short term. Whether political, legal and technological questions can be resolved will largely depend upon how well the Review Conference is able to deal with issues of definition and arrive at agreed interpretations.

If the ABM Treaty is to curb the nature and pace of Soviet and US BMD programmes, the parties must find a consensus on the definition of 'development'. Currently, the Treaty does not contain such a definition. Where does research end and development begin? Article IV allows development and testing of fixed land-based systems and components; Article V prohibits the development and testing of sea-based, air-based, space-based, or mobile land-based systems or components. The United States has argued that laboratory testing is a necessary part of research and does therefore not fall within Treaty restrictions. Testing of prototypes and breadboard models that can be observed by national technical means is prohibited. There is some evidence pointing to an evolving Soviet position on laboratory testing. In his evidence before the US Congress in January 1987, Evgeniy Velikhov, vice-president of the Soviet Academy of Sciences, reportedly stated that the Soviet view on laboratory testing would allow field testing of weapons designed not to perform at full operational power.[20] If Velikhov's statement does indeed reflect the position of the Soviet leadership, the Soviets appear to be moving towards the US understanding of laboratory testing. Under the US understanding, the SDI organization has planned one test designed to 'demonstrate the feasibility of high power infrared chemical lasers for space based application' and another 'intended to validate the weapon potential of a hypervelocity gun and associated miniature kill vehicle technology'.[21] If the Soviets accept these tests as a permitted part of laboratory testing, the ban on testing of space-based ABM systems and components in Article V (1) and the ban on automatic and semi-automatic systems in Article V (2) could be adversely affected.

It is important to reach a joint definition of development in order to address the problem of defining a 'component'. If ABM components can be developed because they can be defined as 'adjuncts', the definition of development is not tight enough. As it stands now, the Treaty does not offer a definition of 'component'. Emerging BMD technologies have begun to produce complex designs for multi-layered BMD systems. Article II mentions ABM interceptor missiles, ABM launchers and ABM radars. The new US interpretation on the applicability of Article II to 'systems based on other physical principles' aside, fact is, that even the technology available for systems and components permitted under the ABM Treaty has advanced a great deal. This raises three important issues. One, if a component can no longer be clearly identified as a component but as a composition of sub-components and adjuncts, the Treaty language becomes vague and allows the parties to develop and test sub-components and adjuncts since in themselves these are neither ABM interceptor missiles, launchers or radars nor can they substitute for ABM interceptor missiles, launchers or radars. Two, sub-components and adjuncts can also be used to enhance the capability of permitted systems and components. Does this make them components also? It is important to resolve this issue soon since it involves, for example, the question of future satellite-based sensor capability. Without an ASAT agreement, ASAT technology, although similar to BMD technology, can function as an ABM adjunct. Three, Article V(1) raises problems for ground-based laser systems. They can be tested at agreed ranges but since they would operate in conjunction with a series of space-based mirrors, the mirrors would be vital for boost-phase defence. This makes the mirrors an essential component of ground-based lasers, yet because they are space-based, their development, testing and deployment is prohibited; in order to arrive at an effective definition of 'component' and 'adjunct', the question of operational interaction between permitted fixed ground-based systems and components and air- and space-based components must be explored.

VI. The ABM Treaty Review Conference

The ABM Treaty limits the purposes to which defensive technology can be put. It codifies the realities of the nuclear age and provides a framework within which Treaty-related issues can be discussed by the signatories. This is what the Treaty *can* do. What it cannot do is control US and Soviet motives in the defensive field.

While offensive strategic nuclear forces still define the strategic-military relationship between the United States and the Soviet Union as one of deterrence, the political relationship between the two countries is defined more by unilateral than co-operative concerns. SDI is a fundamentally unco-operative approach to security. At the strategic level, it ignores the technical fact that mutual assured survival depends on the maintenance of an effective deterrent force by both sides. At the political level, it challenges the Soviets to expand their own version of strategic defence and increase the effectiveness of their offensive nuclear forces. No meaningful agreement on offensive force limitation can be concluded as long as the question of ballistic missile defence undermines deterrence.

There is no technological solution to the harsh realities of the nuclear age. Technology itself is not anathema to deterrence but political will and judgement are required not to let technology alone determine the course of policy.

Both the United States and the Soviet Union seem to have lost faith in nuclear deterrence. President Reagan's SDI has the intention to render nuclear weapons 'impotent and obsolete'. General Secretary

Gorbachev calls for the elimination of nuclear weapons by the year 2000. Are such aims attainable or even desirable? Nuclear weapons cannot be rendered ineffective or swept away. Neither side could have confidence in the effectiveness of its defences or in the good will of the other side not to produce nuclear weapons in secret. Again, the facts of the nuclear age dictate that mutual security must be found through effectively controlling—not merely managing—offensive systems and severely limiting defensive weapons. What does this mean for the upcoming ABM Treaty Review Conference?

Clearly, the ABM Treaty is in deep trouble—so is the entire arms control regime, offensive and defensive. The political, legal, technological and definitional problems are complex and the task of devising a Review Conference agenda will be a challenging task in itself. It is also clear that the Conference must not be over-burdened; in the present situation it is quite likely that the Conference will fail due to the sheer complexity of the issues and the highly-charged atmosphere surrounding it.

Senator Nunn's recent review of the Senate ABM negotiation record and the Senate proceedings strongly criticized the Reagan Administration's legal basis for a broad reading of the ABM Treaty.[22] As the chairman of the Senate Armed Services Committee and long-standing congressional expert on defence issues, Nunn's efforts to hold the Administration to the traditional Treaty interpretation carry political weight. He maintains that his research shows conclusively that the 1972 Senate approved the narrow Treaty interpretation which bans development and testing of space-based BMD technology. Although the Administration has not altered its view on the legal correctness of the broad reading, the Nunn statements have considerably weakened the Administration's case and increased the political heat of the broad versus narrow interpretation debate.

In order for any progress to be made at the Conference, critics of the Treaty should remind themselves that an abrogation of the ABM Treaty gets rid not only of the Treaty but of the arms control regime *per se*. What do the critics have to put in its place? Military superiority, perhaps bought at the price of an offensive/defensive arms race? Even the lesser aim of strengthening deterrence by defending ICBMs is fraught with basic problems. The supporters of the ABM Treaty, on the other hand, must come to realize that as it stands, the Treaty will most probably not be able to survive its own ambiguities, the pace of technological innovation and the political and military opportunism of its opponents. The Treaty will *have* to be amended if it is to survive. It is not sacrosanct, just crucial.

The basic purpose of Treaty amendments must be that of *strengthening* its central provisions: not to deploy ABM systems for nation-wide defence and not to provide a base for such a defence; and the ban on development, testing and deployment of ABM systems or components that are sea-based, air-based, space-based or mobile land-based. It is these central provisions that require political reaffirmation. While that is lacking, problems of definition and technology cannot be resolved.

Notes and references

1   The official ABM Treaty text is printed in the appendix.
2   *President's Commission on Strategic Forces* (US Government Printing Office: Washington, DC, 1983), p. 5.
3   *ABM Treaty Interpretation Dispute*, Hearing before the Subcommittee on Arms Control International Security and Science of the Committee on Foreign Affairs, House of Representatives, Ninety-Ninth Congress, 1st Session, 22 Oct. 1985 (US Government Printing Office: Washington, DC, 1986), p. 3.
4   Ibid.
5   Ibid., pp. 9-18.
6   *The Arms Control Reporter 1986*, Institute for Defense and Disarmament Studies, Brookline, MA, p. 607.B.94.
7   Ibid., p. 607.B.111.
8   *The Arms Control Reporter 1985* (see note 6), p. 575
9   *The Arms Control Reporter 1986* (note 6), p. 575.B.176.
10 Jasani, B., 'Military use of space', *SIPRI Yearbook 1987: World Armaments and Disarmament* (Oxford University Press: Oxford, 1987), pp. 70-1.
11 *Soviet Military Power 1987* (US Government Printing Office: Washington, DC, 1987), pp. 46-7.
12 Ibid., p. 49.
13 *Soviet Strategic Force Developments*, Joint Hearing before the Subcommittee on Strategic and Theater Nuclear Forces of the Committee on Armed Services and the Subcommittee on Defense of the Committee on Appropriations, United States Senate, Ninety-Ninth Congress, 1st Session, 26 June 1985 (US Government Printing Office: Washington, DC, 1986), p. 15.
14 Department of Defense, *The Soviet Strategic Defense Program* (US Government Printing Office: Washington, DC, 1987), p.5.
15 Holloway, D., *The Soviet Union and the Arms Race*, 2nd edition (Yale University Press: New Haven and London, 1984), p. 54.

16   Congress of the United States, Congressional Budget Office, 'Forgoing SALT: Potential costs and effects on strategic capabilities', staff working paper, Aug. 1986, pp. 14-15; see also *Soviet Strategic Force Developments* (note 13), pp. 5-15.

17   MIRV = multiple independently targetable re-entry vehicle.

18   'SDI: progress and challenges', staff report submitted to Senator William Broxmire, Senator J. Bennett Johnston and Senator Lawton Chiles, 17 Mar. 1986, Congress of the United States, 1986, p. 8.

19   Wörner, M., 'A missile defense for NATO Europe', *Strategic Review*, Winter 1986, pp. 13-20.

20   'Soviet Academician makes rare appearance at US Congress', *Nature*, vol. 325 (29 Jan. 1987), p. 381.

21   *Aerospace Daily*, 3 Apr. 1987, p. 22.

22   *Congressional Record, Proceedings and Debates of the 100th Congress*, 1st Session, vol. 133, no. 38 (11 Mar. 1987), no. 39 (12 Mar. 1987), and no. 40 (13 Mar. 1987).

Part II. Background to the current debate

Paper 1. A historical perspective

John W. Finney[1]
5275 Watson Street NW, Washington, DC, 20016, USA

I. The never-ending debate

It came to be known as The Great ABM Debate. In the annals of US politics, there had never been anything quite like it. For three years from 1967 to 1970, a group of senators dared to challenge a weapon proposed by the President—the commander-in-chief—and endorsed by the military. And they almost won. Or perhaps ultimately they did win, for the United States eventually decided not to deploy an anti-ballistic missile defence system.

The debate led to a treaty in 1972 restricting ABM systems, which at the time at least seemed like a historic step by the superpowers. There is still debate over the cause-and-effect relationship between the outcome of the debate and the agreement upon the Treaty, whether the defeat of the ABM critics gave the President the bargaining chip he needed to win Soviet agreement to a treaty or whether the education and enlightenment in nuclear doctrine provided by the debate convinced both sides—and the Soviets in particular—that it was in their mutual best interest to limit defensive weapons in the nuclear age.

In retrospect, the debate had an unexpected, paradoxical effect that has come back to haunt us nearly 20 years later. In ways nobody quite anticipated, it planted the seeds for a reversal by the two sides in the never-ending debate between offence and defence. The United States started off expounding the concept of mutual deterrence based on offensive weapons. But in trying to sell an ABM system, the Nixon Administration was to introduce the threat, which was to become a precept, that the Soviet Union was attempting to upset the deterrent balance by acquiring a first-strike capability against US intercontinental missiles. Out of this argument grew the concept that the United States could not rely entirely upon deterrence by retaliatory weapons but needed the additional deterrent effect of weapons capable of countering Soviet missiles. That was an unsettling, uncomfortable equation. So logically—or illogically, depending on one's point of view—the argument resurfaced that the best way to protect the retaliatory weapons against their presumed vulnerability was to defend them. The door which presumably had been closed shut by the ABM Treaty was thus re-opened for re-introduction of defensive weapons, and the theoretical stage was set, often by the same scientists, strategists and military men who had espoused the original ABM system, for President Reagan's concept of a world where the superpowers could abolish offensive weapons and rely upon defensive ones.

The Soviets, whether for reasons of doctrinal conversion or a sense of technological inferiority or a combination of the two, would seem to have gone through the reverse process in their thinking. They started off saying that offensive systems were immoral and defensive systems moral and came to accept the ABM treaty concept that defensive systems could be destabilizing. Now they are using the logic followed by the United States in selling the ABM Treaty to oppose the strategic concept of a space-based defence system advocated by President Reagan.

There are eerie echoes of the ABM debate in the current controversy over President Reagan's concept, which for want of a better name has been labeled the Strategic Defense Initiative (SDI).[2]

In the discussion on SDI—in the US it has not yet assumed the proportions of a controversy and it is hardly yet a debate—the same technical and doctrinal arguments are surfacing that dominated the ABM debate. Can a leak-proof defence be devised to protect the population or could not the attacker overwhelm it by adding more warheads and mixing in decoys? What assurance would there be that such a complex system, including its computers, would work as designed? How does one phase in such a system without in the process creating uncertainties in the deterrent balance? If the first step is to protect missile sites rather than the general population, would this not give rise to fears that the United States in the transition phase was trying to get itself in a position where it could launch a pre-emptive first strike? And if one is worried about the vulnerability of the missiles, would it not be better and cheaper to build a simpler site-defence system to protect the missiles or to make the missiles mobile? From the standpoint of cost, would it not always be cheaper for the other side to build weapons to fool or overwhelm the defence? Finally, why can't we continue to rely upon a deterrent system that somehow for 40 years seems to have prevented a nuclear war?

On and on the arguments go, just as they did in the ABM debate until the senatorial heads and togas were reeling in confusion. The arguments were not conclusively resolved then, nor are they likely to be now. Debates in Congress or between Congress and the White House are never that precise nor the results clear-cut. Nor do there seem to be the political ingredients for a full-fledged congressional debate on SDI as there was on ABM. The debate of 1967-70 developed during the latter stages of the Viet Nam War, when there were growing misgivings and distrust of the military or military solutions. Under the concept that the Senate was composed of 100 co-equal members, the hierarchical system of seniority was breaking down, which meant that such lordly figures as Senator Richard Russell of Georgia, the Chairman of the Senate Armed Services Committee and an ABM supporter, could no longer impose their judgement upon the Senate. Now the Senate has become so individualistic, it is hard to believe that a significant group could coalesce to challenge a president. Nor, short of a fiscal or economic crisis, does Congress seem willing to face up the question of national priorities, which was always an underlying issue in the ABM debate.

It would be nice to believe that the ABM debate was fought out in grand terms about the proper nuclear doctrine for the United States and the world. To an extent it was, although there was always a question of how many senators ever really understood the intertwined doctrinal and technical issues that only seemed to become more arcane as the debate proceeded. But domestic political considerations were undoubtedly more important than strategic policy in first instigating an ABM system and then in shaping the debate over whether it should be built. And in the end political personalities and loyalties were more important than strategic policy in determining the outcome of the debate.

II. McNamara's case for and against an ABM system

A convenient place to pick up on the debate is a speech on 18 September 1967 on 'The dynamics of nuclear strategy' by Defense Secretary Robert S. McNamara. It was a speech remarkable for its logic and apparent contradiction in logic. On the one hand, in a cogently reasoned argument, he described why it would be foolish and feckless to build an ABM defence against the Soviet Union; and then, in a clashing shift of gears, McNamara presented the reasons—'marginal', as he described them—for building a 'light' ABM system to defend against a Chinese missile threat with the 'concurrent benefit' of providing 'a further defense of our Minuteman sites against Soviet attack'.

Ironically, a speech that started the United States down the road toward building an ABM system was later to supply the critics with the basic arguments on why such a system should be dismantled.

On the question of whether the Soviet Union was acquiring nuclear superiority—a xenophobic question that was to dominate and constantly confuse the ABM debate—the McNamara speech contained the startling admission that the United States had built more strategic warheads than it had planned or required. As a result, the United States had nuclear 'superiority', but such a term was of 'limited significance' in the nuclear equation. As he explained the dynamics of the nuclear arms race:

We both have strategic arsenals greatly in excess of a credible assured destruction capability. These arsenals have reached that point of excess in each case for precisely the same reason: We each have reacted to the other's buildup with very conservative calculations. We have, that is, each built a greater arsenal than either of us needed for a second strike capability, simply because we each wanted to be able to cope with the 'worst plausible case'.

On the question of whether the Soviet Union was attempting to acquire a first-strike capability against the United States—a contention that was to become one of the principal arguments of the Nixon Administration for building an ABM—McNamara said he did not believe so, but in any event the question was irrelevant.

We do not possess first-strike capability against the Soviet Union for precisely the same reason that they do not possess it against us. And that is that we have both built up our 'second-strike capability' to the point that a first-strike capability on either side has become unattainable.

In response to assertions that the US should match the Soviet Union, which was deploying an ABM system around Moscow—the keep-ahead-of-the-Russians argument that was so popular in Congress in the post-Sputnik period—Mr McNamara said that any present or foreseeable ABM system 'can rather obviously be defeated by an enemy simply sending more offensive warheads, or dummy warheads, than there are defensive missiles capable of disposing of them'. If the United States were to deploy an anti-Soviet ABM system, the Soviets would respond by increasing their offensive capability. This would 'trigger a senseless spiral upward of nuclear arms' and in the end both sides would 'be relatively at the same point of balance on the security scale that we are now'.

Having thus so cogently presented the case for not building an ABM system, the same argument he had briefly presented to the Soviets at the Johnson-Kosygin meeting in Glassboro earlier in the year, Mr McNamara concluded the speech with what appeared to be an amazing flip-flop in logic. An assured destruction capability was sufficient to deter the Soviets but not the Chinese. By the mid-1970s China might have a 'modest force' of intercontinental missiles and 'become so incautious as to attempt a nuclear attack on the United States'. Therefore, he concluded, 'there are marginal grounds for concluding that a light deployment of US ABMs against this possibility is prudent'. But he coupled this decision with a warning that was to be prophetic:

The danger in deploying this relatively light and reliable Chinese-oriented ABM system is going to be that pressures will develop to expand it into a heavy Soviet-oriented ABM system. We must resist that temptation firmly, not because we can for a moment afford to relax our vigilance against a possible Soviet first strike but precisely because our greatest deterrent against such a strike is not a massive, costly, but highly penetrable ABM shield but rather a credible offensive assured destruction capability.

Mr McNamara's contradiction is not so baffling if viewed in a political context. Indeed, the decision to proceed with an ostensibly anti-Chinese ABM system is a revealing illustration of how domestic political considerations were at least as important as strategic arguments in starting the United States down the road to an ABM system. The pressures had been mounting in both Congress and the Joint Chiefs of Staff to deploy an ABM system, which had been under development for years. In his budget presented to Congress in January 1967, President Johnson, in a deal he struck with Mr McNamara, sought to forestall those pressures by including money to buy components for an ABM system, but stipulating they would not be deployed pending an attempt to enter into arms control negotiations with the Soviet Union. By late 1967, with a presidential election year in the offing, the political pressures were becoming irresistible. The Glassboro summit had not led to arms control negotiations. Republicans were beginning to make sounds about raising a 'defense gap' issue in the 1968 elections, and President Johnson, who as a senator had helped create the 'bomber gap' and 'missile gap' issues that bedeviled the Republicans, fully appreciated how effective a 'gap' could be as a political device. But not only the Republicans were concerned. Democrats on the Joint Congressional Committee on Atomic Energy—and Senator Henry M.

Jackson in particular—were keeping up a drumfire of warnings about Soviet and Chinese advances in nuclear weapons. The anti-Chinese system, therefore, became a fall-back position developed by McNamara and accepted by President Johnson, who in the midst of the Viet Nam War wanted to maintain a semblance of unity in his official family.

III. ABMs become a nation-wide issue

Normally the Administration's request for construction funds for an ABM system would have sailed through Congress without much debate. But there were two developments the Administration had not anticipated. Both illustrated how individuals were to influence an abstract debate. One involved William G. Miller, a 37-year old former Foreign Service officer who had left the State Department and joined the staff of Senator John Sherman Cooper, Republican of Kentucky. Mr Miller was typical of a group of bright, young, assertive individuals who had left the executive branch seeking greater power and influence in Congress and who behind the scenes were to shape the ABM debate. In spring 1968, when the Senate was considering a military authorization bill, Mr Miller walked down a side aisle in the Senate chamber and handed Senator Cooper a one-page memorandum suggesting that the matter of funds for the Sentinel ABM system 'deserves questioning' because it had not been 'fully proven by research'. Senator Cooper harumphed in agreement, and with that the ABM debate was launched. Senator Cooper joined forces with Senator Philip Hart, Democrat of Michigan, a fortuitous combination since both commanded respect among liberals and conservatives alike in the Senate, and their sponsorship provided respectability for the uncommon challenge to the military-congressional complex that had long had its own way on defence matters.

The second unexpected development was a grass roots protest that erupted against the planned sites for the Sentinel system. Its 15 sites were to be located near major cities, which were to be defended by short-range Sprint missiles against warheads that the long-range Spartan missiles had missed; and the missiles would all be armed with nuclear warheads. For years the Army had stationed Nike Hercules anti-aircraft missiles tipped with nuclear warheads around cities with no public protest because the Army never acknowledged the presence of the nuclear weapons. But with Sentinel it was no longer a secret. From Boston to Honolulu city councils, church groups, peace groups, conservationists, union leaders, real estate developers and scientists banded together to protest the planned emplacement of nuclear-tipped missiles in the backyards of suburban America. Nothing makes a congressman think quite as much as an organized, grass roots protest from the constituents back home. In an unanticipated way, the anti-ABM effort was given a political momentum.

If private citizens were aroused so were members of the scientific community. Not since the great debate in 1946 over the atomic energy legislation had scientists become so involved in a political issue in Washington. The scientists included: George Kistiakovsky of Harvard; James Killian and Jerome Weisner of MIT, all of whom had served as science advisers to presidents; Jack Ruina and George Rathjens of MIT, who had worked in the Pentagon; and David Inglis of the Argonne National Laboratory. Prowling the halls as a lobbyist was Jeremy Stone of the Federation of American Scientists, the remnant of the scientific group that in 1946 successfully challenged military control of the atom. But most dominant in the debate was Wolfgang Panovsky of Stanford, who with thick glasses and heavy accent fitted the senators' image of a scientist and who day after day lead the senators through the intricacies of nuclear warfare in private briefings and public hearings. The unusual tutoring schooled the senators in the technical limitations and weaknesses of the Sentinel system. Since they were better informed, the anti-ABM senators succeeded in putting on the defensive the Senate Armed Services Committee, which over the years had grown accustomed to not having its judgement challenged on military matters.

The anti-ABM forces, distinctly in the minority, started off cautiously and uncertainly. In June 1968 the Senate rejected by a 52-34 vote a Cooper-Hart amendment to delay the construction funds for a year. It was surprising that so many senators were willing to vote against a weapon system wanted by the President and the military. The anti-ABM forces were obviously picking up support, so much so that the incoming Nixon Administration decided to call a halt in deployment of the Sentinel system and to re-examine the ABM programme.

IV. From population defence to point defence—Nixon and the Soviet threat

Shortly after taking office the Nixon Administration reoriented the ABM programme away from a 'thin' area system designed to protect populations against a potential Chinese attack to a point defence designed to protect Minuteman missiles in their silos against a pre-emptive Soviet attack. The name was changed to Safeguard. The justifications offered for the shift in emphasis reflected a significant if evolutionary change in nuclear doctrine. The primary objective, President Nixon explained, must be 'protecting our deterrent', which was the 'best preventive for war' but which was becoming 'increasingly vulnerable' to Soviet attack. A defence of Minuteman bases would be less provocative to the Soviet Union since the reoriented ABM system would be 'so clearly defensive in character'. With the thin, city-defence system, Mr Nixon explained, there was always the possibility that it would be expanded into a 'heavy' system, and this 'tends to be more provocative in terms of making credible a first-strike capability against the Soviet Union'. Then, in words that may be recalled someday in the debate over President Reagan's Strategic Defense Initiative, Mr Nixon declared 'there is no way that we can adequately defend our cities without an unacceptable loss of life'. In explaining the difference between trying to protect missile bases and cities, the President said:

When you are looking toward a city defense, it needs to be a perfect or near perfect system because, as I examined the possibility of even a thick defense of cities, I have found that even the most optimistic projections, considering the highest development of the art, would mean that we would still lose 30 million to 40 million lives. That would be less than half of what we would otherwise lose. But we would still lose 30 million to 40 million.

When you are talking about protecting your deterrent, it need not be perfect. It is necessary only to protect enough of the deterrent that the retaliatory second strike will be of such magnitude that the enemy would think twice before launching a first strike.

Politically the reorientation of the ABM programme was a shrewd move. By moving the ABM sites away from the cities to two remote locations initially in Montana and North Dakota, the Administration succeeded in defusing the grass-roots protest that was engulfing Congress. Given the post-Sputnik fear of the Soviets, it was much easier to sell an anti-Soviet system than one designed against some hypothetical Chinese threat.

It was an approach immediately seized upon by Defense Secretary Melvin Laird, the former congressman who undoubtedly was the craftiest politician ever to occupy the post of Secretary of Defense. Pointing to the large SS-9 missiles that the Soviets were deploying, Laird warned: 'They are going for our missiles, and they are going for a first-strike capability. There is no question about that.' Of course, there were a lot of questions about that, and Senator J. William Fulbright, the Chairman of the Senate Foreign Relations Committee accused Laird of using a 'technique of fear' to 'sell' the ABM programme. But Laird succeeded in implanting the fear that the Soviets were acquiring large, accurate missiles that left US retaliatory missiles vulnerable to attack, a fear that has dominated and influenced US strategic thinking to this day.

In a subtle way not completely appreciated at the time, it also marked a shift away from the doctrine of Mutual Assured Destruction. The theory or the threat, depending upon one's point of view, was that the Soviet Union might conduct a pre-emptive, first strike against the US land-based retaliatory missiles. This would leave the United States no alternative but to attack Soviet cities in retaliation, knowing that the Soviets still had the ability to attack US cities. This re-examination of strategic doctrine found formal expression in President Nixon's State of the World message in 1970. The President accepted the strategic goal of 'sufficiency', but he then went on to ask:

...the growing strategic forces on both sides pose new and disturbing problems. Should a president, in the event of a nuclear attack, be left with the single option of ordering the mass destruction of enemy civilians in the face of the certainty that it would be followed by the mass slaughter of Americans? Should the concept of assured destruction be narrowly defined and should it be the only measure of our ability to deter the variety of threats we may face?

Out of these questions grew the strategic concept that the United States had to reduce the vulnerability of its land-based missiles, either by making them mobile or defending them, and that as a deterrent against a Soviet first strike the United States needed its own counterforce weapons (such as the MX missile) capable of attacking Soviet missiles. It also led to the negotiating position of the Nixon Administration that in any arms control agreement there would have to be limitations on Soviet missiles and that a defensive ABM system could be a bargaining chip in obtaining control over offensive weapons.

The ABM debate was to spawn a whole new generation of strategic thinkers who were the nuclear-age counterparts to the Talmudic and Jesuit scholars of earlier eras—but the net effect was to raise questions whether mutual deterrence was a stable, satisfactory balance. From there it was not too big a jump to President Reagan's conclusion that deterrence should be based on defensive rather than offensive weapons.

V. Safeguard faces a technical challenge

The Nixon Administration's reorientation of the ABM programme only served to fuel rather than subdue the ABM debate in the Senate. Undoubtedly the Administration thought it could carry the day by raising the Soviet threat; what it apparently did not anticipate was that with the reorientation the ABM system was open to technical challenge by the ABM critics in the Senate. By then they had been well-tutored in the intricacies of an ABM system, far better than the defenders from the Senate Armed Services Committee. The problem was that the Sentinel system, particularly with its missile-aiming radar, had been designed for area defence of populations; the question, immediately seized upon by the opposition, was whether the renamed system could work effectively in providing point defence of Minuteman missile fields. For weeks the Senate became enmeshed in technical debate over whether the Soviet threat was as great and as serious as projected by the Pentagon, whether the Safeguard system would work in defending the Minuteman missiles, whether its radar was vulnerable to attack or blackout and whether the Soviets could not exhaust the supply of defensive Sprint missiles just by adding a few score more warheads to the attack.

While the debate was going on in the Senate, the Air Force was beginning to test multiple, independently targetable warheads, known as MIRVs, for its Minuteman missiles. It was a perfect illustration of the action-reaction cycle that Mr McNamara had warned was fuelling the nuclear arms race. In a way Mr McNamara had fuelled the latest cycle. Rather than permitting the Air Force to build more Minuteman missiles, he had authorized it to develop multiple warheads for the missile as a way of overwhelming an ABM system the Soviets were building around Moscow. Several years later a point of no-return in terms of arms control was being reached. The Defense Department was flight-testing MIRVs which were needed because the Soviet Union was building an ABM system; and the Pentagon wanted an ABM system to defend its Minuteman missiles because it feared the Soviet Union was developing accurate, multiple warheads for its large intercontinental missiles.

The two-year-old debate came to a suspenseful climax on the floor of the Senate on the afternoon of 6 August 1969. By one- and two-vote margins, the initial phase of the Safeguard system was approved. Never had a major weapons programme come so close to being defeated in Congress. But the Administration had won, and construction began on an ABM system.

VI. ABMs as a bargaining chip

That should have been the end of the ABM debate, but in 1970 the Nixon Administration did the unexpected. It proposed the expansion of the Safeguard protection to two other Minuteman sites in Missouri and Wyoming and revived the idea of building a 'light' area defence by proposing preliminary work on three urban sites, including Washington, DC. In what seemed to be a shift in rationale, the President said an area defence against a possible Chinese attack was 'absolutely essential' if the United States was to have a 'credible' foreign policy in the Pacific, and he contended the system would be 'virtually infallible' against a Chinese attack.

It was too much even for the Senate Armed Services Committee, which ruled out a start on an anti-Chinese system but approved expansion of Safeguard to two additional Minuteman sites. By then, however, the ABM opposition was on the defensive. The strategic arms limitation talks (SALT) with the Soviet Union had started, first in Helsinki and then in Vienna, and the opposition did not want to get itself in the position of seeming to undercut the President at a time when the talks were showing some promise of controlling defensive as well as offensive weapons. So rather than proposing abolition of the Safeguard system, as its past argumentation would have dictated, the opposition decided to draw the line at limiting the system to the two sites already approved.

What ensued was a political and geopolitical poker game masterfully played by the Nixon Administration against the Soviets in Vienna and the Senate critics back home. The technical arguments became submerged in a debate over whether expansion of the Safeguard system was needed as a bargaining chip to induce the Soviets to agree to a limitation on offensive as well as defensive strategic weapons. In effect, the Administration was raising the ante and trying to get the Soviets to lay offensive weapons on the negotiating table by threatening an expansion of the Safeguard system. The Administration's rationale—and its appraisal of the Soviet negotiating stance—was best explained by Senator Jacob K. Javits, Republican of New York, who, obviously on the basis of a briefing by the Administration, told the Senate:

It has been explained to me that the Soviet negotiating team represents a coalition of interests having diverse reasons for wanting a SALT agreement. It is said that the Soviet negotiating coalition is a delicately constructed one and that the element representing the military is the most reluctant and suspicious element. The group representing the Soviet Union's military viewpoint is said to be interested primarily in halting the development of an American ABM system. Presumably—using the 'worst case' war gaming approach—the Soviet strategic planners place a higher efficiency factor on Safeguard's capabilities than our own scientific community does. Accordingly, it is contended that the Soviet military component, which is prominently represented in the Soviet negotiating team, might lose interest in achieving a SALT agreement if the Safeguard system is killed off in the Senate. The defection of the Soviet military element could disrupt the delicately constructed Soviet negotiating consensus and thus jeopardize an agreement otherwise desired by other elements of the Soviet hierarchy.

The bargaining chip argument was also used effectively in the Senate to win expansion of the Safeguard system. Repeatedly the Senate was told that failure to expand the system would lead to failure in the strategic arms talks—a responsibility that a majority of senators did not want to assume. The Cooper-Hart amendment to restrict the system to two sites failed by a 52-47 vote. A shift of three votes would have reversed the outcome, and how the Administration obtained those three votes illustrates that despite all the technical and geopolitical argumentation the personal element was to decide the debate. Two of the votes were obtained with a one-paragraph telegram sent by Gerard C. Smith, the head of the US delegation to the SALT talks, on instructions from the White House. It was in those days an unusual intervention by an ambassador in a Senate debate, and the telegram was circulated privately only among wavering and uncommitted senators. The telegram disclaimed any knowledge of amendments pending before the Senate but went on to express concern about the impact of the Senate debate upon the SALT negotiations and to urge that the United States not assume a static position on Safeguard deployment. The telegram was credited with shifting the votes of two Senators—Thomas McIntyre of New Hampshire and James Pearson of Kansas—who had opposed the Safeguard system in the previous year.

How the third vote was obtained is the poignant story of one of the final acts of an ageing, failing senator who had once been one of the great figures in the Senate. On the evening preceding the vote, Senator Clinton P. Anderson of New Mexico, who had voted for Safeguard in the previous year, issued a statement announcing he would vote against expansion because he believed the United States should proceed in a 'cautious' way in deploying an ABM system. When Senator Henry M. Jackson, a close Anderson friend, and Mrs Anderson heard about the press release, they demanded that it be recalled, only to be told that it was too late because the statement had already been given to the *New York Times* and the wire services. The next day, Senator Anderson was seated at his desk near the front of the Senate chamber as the roll was called on the Cooper-Hart amendment. 'Anderson', the clerk called. The senator opened his mouth but such was his frailty that his reply was unintelligible. 'Anderson', the clerk called out a second time. Again the reply was indistinct. On the third call, Senator Jackson, who was away from his assigned desk and seated with his arm around Senator Anderson, voted for his colleague by saying 'the senator from New Mexico votes no'. In Senator Anderson's office some wept at what they regarded as humiliation of their boss on the floor of the US Senate where his voice had once echoed with power.

Thus did the Nixon Administration win the 'bargaining chip' to play against the Soviets. The political maxim was firmly implanted that the only way to control arms is to build arms. Since then it has been claimed—and nobody can disprove it—that the Senate vote led to an ABM Treaty and an agreement limiting offensive weapons.

VII. The interpretation issue—what did the Senate believe it was approving?

There was to be a fascinating postscript to the great ABM debate some 15 years later, with the announcement of a 'new interpretation' of the Treaty. The ABM supporters then found their earlier questions about the Treaty being used to challenge the Reagan Administration's interpretation that the Treaty permitted development of a space-based defence system.

By the time the ABM Treaty reached the Senate in the summer of 1972, the debate had dissipated. The opponents thought that they had achieved a victory in a treaty that limited an ABM system to two sites and thus precluded deployment of a nation-wide system. And the proponents were not about to challenge a popular treaty that seemed to curb the atomic arms race. The Senate gave its approval to ratification of the Treaty by a vote of 88 to 2, a vote so one-sided that it indicates the lack of controversy over the Treaty.

It would be nice to think that the senators knew in detail what they were approving, but that is not in the nature of senators when faced with what they regard as a non-controversial measure. In fact, it is probably fair to say that only a half dozen senators fully understood the provisions of the Treaty dealing with future development of ABM systems, and they were mostly senators, such as Jackson, who had supported the ABM programme. Their concern at the time was that down the road the Treaty might preclude development or deployment of futuristic ABM systems using such exotic technologies as lasers and/or based in space rather than on the ground.

Such was the overwhelming sentiment in favour of the Treaty that nobody in the Senate or in the press paid much attention at the time to what were regarded as the nit-picking questions of Senator Jackson and a few others or to the responses of the Administration. The exchanges might have languished in the dust of history if the State Department's legal office had not come up in 1985 with a new interpretation of the ABM Treaty, asserting that it permitted a go-ahead on the Strategic Defense Initiative. The reinterpretation was said to be based in part on a reading of the record of the Senate consideration of the Treaty. That prompted Senator Sam Nunn, who was not around when the ABM debate took place, to delve into the archives of the Senate Armed Services Committee and come up with a remarkable piece of historical research to challenge the Administration's reinterpretation that the Treaty would permit testing of futuristic components, such as lasers, for a space-based defensive system. In a speech on the Senate floor,[3] Senator Nunn pointed out that during testimony on the Treaty in 1972 Administration officials drew a distinction between mobile, space-based systems and fixed, land-based systems when it came to the development and testing of new technologies, such as lasers. Under Article V of the Treaty, Administration officials testified, it would be permissible to develop and test—but not deploy—fixed, land-based systems using new technologies. As for sea-based, air-based, space-based or mobile land-based systems using new or exotic technologies, there could be no development or testing beyond the laboratory stage.

It was to provide a historical irony for an ABM debate that was not in the personal memory of most of the senators. Questions asked then by ABM supporters were to produce testimony on how under the Treaty the United States could not advance beyond the laboratory stage in the development and testing of a space-based ABM system that a new generation of ABM supporters now advocate.

VIII. Epilogue

What lessons for the future can be drawn from the ABM debate, particularly as the two superpowers are once again negotiating the future of defensive weapons? Perhaps the following:

1. A president—whether it is Mr Reagan or his successor—does not automatically face concerted congressional opposition. The ABM debate was one of those rare moments of spontaneous congressional combustion which require an unusual concatenation of conditions. There was a mood of disenchantment then with the military whereas the political temper is now chauvinistic. A large group of senators was willing to coalesce behind the leadership of a couple of their respected colleagues, something that is hard to envisage in the individualistic Senate of today. The moderate Republicans who teamed up with Democrats to provide a near-majority are a dying breed. Democrats, meanwhile, are in disarray on defence policy. The Senate Foreign Relations Committee, which provided the core of the ABM opposition, is in decline, while the Senate Armed Services Committee is rising once again in influence. The World War II generation of scientists, who were treated almost as oracles in the post-Sputnik period, is dying off, to be replaced by a new generation of scientists which commands neither the public reverence of the past nor seems interested in political activism. And to a far greater degree than 20 years ago, we live in a television age, which gives the upper hand to the executive branch in shaping public opinion.

If there is one factor that may ignite a congressional fight over the SDI programme it is the budget. The question of national priorities between military and domestic needs was always an underlying theme in the ABM debate, although it never quite came to the fore. But then we were talking about a weapon programme costing tens of billions of dollars, and deficits were running at $10 billion and less. Now we are talking about a weapon concept that will cost hundreds and hundreds of billions of dollars, and the deficit is running at nearly $200 billion a year. The ingredients are there for Congress to take out its frustration over the deficit on the SDI programme.

2. A presidential invocation of national security, combined with a plea for a bargaining chip to deal with the Soviets, will, almost inevitably, carry the day in Congress. While President Nixon may have accepted the concept of 'sufficiency', superiority remains the prevailing political term—either as a goal for the United States or something the

Soviet Union is about to achieve. That is probably truer now than it was 20 years ago.

3. In a way that the critics never quite appreciated at the time, the outcome of the ABM debate demonstrated a linkage between offensive and defensive arms when it comes to negotiations with the Soviet Union. It is a concept that the Nixon Administration understood from the outset and which the Soviets seem to have come to accept. Whether the linkage is understood and accepted by the Reagan Administration remains to be seen. SDI, which is of obvious concern to the Soviets, could still provide the leverage for Soviet agreement to restrict the large missiles that concern the United States. But this presumes that SDI becomes a bargaining chip and not a vision. And in a curious throwback to the past, a compromise may well depend upon a renewed agreement upon the terms of the treaty that grew out of the ABM debate.

Notes and references

[1]   John W. Finney is a former reporter and news editor in the Washington Bureau of the *New York Times* who covered the ABM debate.

[2]   As was seen in the ABM debate, names can play an important role in shaping opinion. The ABM system started off in the Johnson Administration as Sentinel and then was changed by the Nixon Administration to Safeguard—two benign names that would belie any hostile intent. Conversely, little did its advocates appreciate that in an age of acronyms, mutual assured destruction would be reduced to MAD, giving the critics a handle to implant in the public's mind—and perhaps a president's—the suspicion that reliance upon offensive weapons for deterrence was a mad idea. The lesson here is that the Reagan Administration will have to come up with a better, more catchy name than SDI if it wants to avert the use of the admittedly perjorative name Star Wars that the critics have been able to fix on the Reagan initiative.

[3]   Congressional Record of 11 Mar. 1987.

Paper 2. The Treaty's basic provisions: view of the US negotiator

Gerard C. Smith[1]
The Consultants International Group, Suite 400, 1616 H Street, NW, Washington, DC, 20006, USA

I. Introduction

The Anti-Ballistic Missile (ABM) Treaty of 1972 banned the deployment of nation-wide defences against strategic ballistic missiles by the United States and the Soviet Union. To ban defences may seem paradoxical, until it is realized that defences, in effect, amount to an effort to disarm one's adversary's deterrent forces. The Treaty was based on a mutual realization that the deployment of missile defences would only set off a redoubled arms race: if one side deployed defences, the other side was sure to try to match them, and each side was sure to increase its offensive forces to overcome the other's defences. The ABM Treaty nipped this nascent offence-defence race in the bud. No competition in deployment of missile defences developed, and it seems likely that the competition in offensive arms that did develop would have been more heated had defensive arms not been under international control. By reducing the incentives to build up larger offensive forces, the ABM Treaty made the SALT II limits on offensive arms possible, although the major reductions in offensive forces that were hoped for when the ABM Treaty was signed have not taken place.

In addition to spurring the arms race, deployment of nation-wide defences could destabilize the military balance in a deep crisis. Defences would inevitably be more effective against a ragged, disorganized retaliation than they would against a large, carefully planned first strike. So they might increase each side's incentive to strike first and heighten the risk of war. This was an additional reason to ban the deployment of widespread defences against ballistic missiles.

II. Article I: the basic commitment

The fundamental undertaking of the Treaty is contained in the second paragraph of Article I: 'Each Party undertakes not to deploy ABM systems for a defense of the territory of its country...'. With this one clause, the world's most powerful nations stated their intention to refrain from building widespread ABM defences and thus to remain defenceless against missile attack. That commitment is more significant than any of the Treaty's specific provisions. It reflects a fundamental judgement that deployment of widespread defences against ballistic missiles would touch off an arms competition that would prejudice the security of both countries.

While the ban on deployment of a nation-wide defence is fundamental, it is not enough to cope with the security problems posed by ABMs. The United States believed that a treaty limiting ABM systems to very low levels would not have much significance if one side could abrogate it and rapidly deploy defences before the other side could respond. Many of the Treaty's provisions were intended to meet this 'breakout' concern and undergird Article I's broad prohibition on the deployment of nation-wide defences.

Article I also bans both the deployment of 'regional' defences (except as specifically allowed by Article III), and the creation of a 'base' for a nation-wide defence. The ban on creation of a 'base' would forbid, for example, deployment of a widespread network of ABM-capable radars, even if they were claimed to be for purposes permitted by the Treaty. Radar restraints are discussed in more detail below.

III. Article II: definition of ABM systems and components

Article II defines the ABM systems and components limited by the Treaty:

For the purposes of this Treaty an ABM system is a system to counter strategic ballistic missiles or their elements in flight trajectory, currently consisting of:

(*a*) ABM interceptor missiles, which are interceptor missiles constructed and deployed for an ABM role, or of a type tested in an ABM mode;

(*b*) ABM launchers, which are launchers constructed and deployed for launching ABM interceptor missiles;

(*c*) ABM radars, which are radars constructed and deployed for an ABM role, or of a type tested in an ABM mode.

'ABM system' is defined functionally, as a system 'to counter strategic ballistic missiles or their elements in flight trajectory'. The components of such a system are described as 'currently consisting of' ABM missiles, launchers, and radars. Despite recent arguments by some officials of the current US administration, this language was not intended to exclude other components, based on other technologies such as lasers, which might serve the same roles as ABM missiles, launchers and radars. On the contrary, the word 'currently' was inserted to insure that the definition would not be limited to 'current' components, but would cover future ABMs based on novel physical principles. (A more detailed discussion of this issue is included below, in the section discussing Article V.)

The language describing what types of missiles, launchers and radars should be considered ABM components was a compromise. The United States sought to base the Treaty's limitations primarily on the capabilities of weapons. The Soviets preferred to focus on the parties' intentions in deploying those weapons. Systems ostensibly intended for air defence, they argued, should not be limited, even if they might be thought to have some capability against ballistic missiles. The US could not accept that, since the real intent behind a piece of hardware cannot be objectively verified. In the end, it was agreed that any missile, launcher or radar 'constructed and deployed for an ABM role' or 'tested in an ABM mode' would be included under the Treaty's restraints. Combined with the restraints on giving non-ABM components an ABM capability (contained in Article VI), and the restraints on deployment of non-ABM radars (Article VI and agreed statement F), the Treaty significantly reflects concerns over the possible ABM capabilities of non-ABM components and goes a good way towards establishing capability rather than intent as the criterion for limitation.

The second paragraph of Article II makes clear that the ABM components to be limited by the Treaty include not only those which were operational in 1972, but also those under construction or testing, or those being overhauled, repaired, modernized or mothballed. The Treaty covers them all.

IV. Article III and the Protocol: permitted deployments

Article III describes the limited deployments of ABM systems that are permitted by the Treaty, and bans all other deployments of ABM systems. At the time, the Soviets had an ABM site to defend Moscow and the United States was constructing two sites to defend some missile silos. The Treaty allowed each side two sites, one of the type it was currently deploying plus one of the type being deployed by the other side. A 1974 Protocol reduced the number of allowed sites to one each, which could be either for defence of the national capital or for defence of a missile field.

Article III begins with a provision prohibiting all ABM deployments not explicitly allowed: 'Each party undertakes not to deploy ABM systems or their components except that...'. The only allowed deployments are fixed, land-based ABM launchers, missiles, and radars. Each site may include no more than 100 ABM launchers equipped with 100 ABM missiles. Combined with a ban on rapid-reload and multiple-warhead ABMs in Article V, this drastically limits the capability of the permitted ABM system. One hundred interceptors would be of little help in defending against strategic forces with thousands of warheads. Article III also includes precise limits on ABM radars. They are limited in location (for the national capital site) or in number and capability (for the ICBM defence site).

Article III implicitly bans the deployment of futuristic ABM systems and components, such as those based on lasers or particle beams. It begins by forbidding everything that is not allowed, and then allows only ABM systems based on launchers, interceptors and radars, hence excluding ABMs based on other technologies. Development of futuristic ABMs is permitted only if they are fixed and land-based, and testing is limited to agreed test ranges. The two sides agreed that should such testing lead to what appears to be a workable system, deployment would be forbidden until specific limitations were agreed, analogous to the limits on launchers, missiles and radars in Article III. This is spelled out in agreed statement D, which clarifies Article III's implicit ban on the deployment of futuristic ABMs:

In order to insure fulfillment of the obligation not to deploy ABM systems and their components except as provided in Article III of the Treaty, the Parties agree that in the event ABM systems based on other physical principles and including components capable of substituting for ABM interceptor missiles, ABM launchers, or ABM radars are created in the future, specific limitations on such systems and their components would be subject to discussion in accordance with Article XIII and agreement in accordance with Article XIV of the Treaty.

In short, even if they are fixed and land-based, ABM systems or components 'based on other physical principles' cannot be deployed unless the Treaty is amended to provide 'specific limitations' on them.

V. Article IV: test ranges

Article IV regulates ABM test ranges. Development and testing of all fixed, land-based ABM systems is allowed, but is limited to those test ranges already in existence, or those additionally agreed between the two sides. In addition, Article IV allows a maximum of only 15 ABM launchers for test purposes. With the location of test ranges constrained and only 15 launchers allowed, there is little danger of ABM systems at test ranges contributing significantly to a regional or nation-wide defence, or providing a base for such a defence.

VI. Article V: ban on mobiles and reloadables

Article V prohibits the development, testing and deployment of all sea-based, air-based, space-based, or mobile land-based ABM systems and components. Common understanding C specifies that a 'mobile' component is one that is not 'of a permanent fixed type'. For example, development and testing of an ABM radar in a van that could be rapidly transported from place to place would be prohibited.

Article V's restraints on sea-based, air-based, space-based and mobile land-based ABMs are far more severe than those the Treaty places on fixed, land-based ABMs—and for good reason. Such mobile ABMs are inherently expandable beyond the single allowed site. Hence they would undercut the Treaty's key provisions banning nation-wide or regional defences, or the base for such a defence. Requiring little site preparation, they might be rapidly deployed once produced, raising further 'breakout' concerns. And their mobility would make numerical limits on their deployment inherently more difficult to verify. Keeping such systems forever in the infancy of research was one of the great accomplishments of the ABM Treaty.

Similarly, the second paragraph of Article V bans the development, testing and deployment of launchers capable of launching more than one ABM interceptor missile at a time, or capable of rapid reload. Such rapid-fire launchers would have undercut the limit on ABM fire-power imposed by the Treaty's limit of 100 ABM launchers. Agreed statement E broadens the limitation to include a ban on development, testing and deployment of multiple-warhead ABM missiles, which otherwise would have had the same prejudicial effect.

The limits of Article V begin with the 'development' stage. Research is not limited. But drawing the line between permitted research and forbidden development is a difficult problem. When the Treaty was negotiated, the US believed that the delegations had come to a reasonable understanding as to what stage of the research and development process was to be constrained, although no formal definition of 'development' was reached. In essence, the word 'development' as used in the Treaty refers to the stage that follows research, and begins at the time when prototypes leave the laboratory and are ready for field testing. Limits on earlier stages of the research and development cycle could not be reliably verified. During the Treaty ratification hearings, I was asked to clarify this point, and provided the following written response, which was based on the negotiating history and had been worked out among the Washington agencies involved in SALT:

The obligation not to develop such systems, devices, or warheads would be applicable only to that stage of development which follows laboratory development and testing. The prohibitions on development contained in the ABM Treaty would start at that part of the development process where field testing is initiated on either a prototype or breadboard model. It was understood by both sides that the prohibition on 'development' applies to activities involved after a component moves from the laboratory development and testing stage to the field testing stage, wherever performed. The fact that early stages of the development process, such as laboratory testing, would pose problems for verification by national technical means is an important consideration in reaching this definition. Exchange with the Soviet delegation made clear that this definition is also the Soviet interpretation of the term development...Article V...places no constraints on research and on those aspects of exploratory and advanced development which precede field testing. Engineering development would clearly be prohibited.

VII. The 1985 unilateral US reinterpretation

In October 1985, the Reagan Administration suddenly announced a radical reinterpretation of the ABM Treaty. Under the new interpretation, Article V is read as permitting development and testing of space-based lasers and particle beams, despite the unambiguous language forbidding development, testing and deployment of all mobile ABM 'systems' and 'components'. This unilateral reinterpretation reversed the established understanding of the Treaty. It contradicted the historic understanding accepted by the Treaty's negotiators, including myself; the US Senate, which ratified it; and every Administration since

President Nixon's, including the Reagan Administration before October 1985. It rips the heart out of the Treaty's key provisions.

The announcement of this fundamental change in the ABM Treaty provoked a storm of criticism, leading Secretary of State Shultz to announce that while the Administration believed the new interpretation to be legally correct, it would continue to abide by the traditional view of the Treaty's limits, at least for the time being.

How was this loophole found in the clear language of Article V? The Reagan Administration argues that 'exotic' technologies are not 'systems' and 'components' as defined by the Treaty, and hence are not covered by Article V. Thus the new interpretation drastically alters not only Article V, but much of the rest of the Treaty as well. In the Reagan Administration's view ABM systems based on exotic technologies would be completely unlimited were it not for agreed statement D. Agreed statement D is now seen as a substantial amend-ment to the Treaty, forbidding deployment of exotic systems that would otherwise be allowed.

The question boils down to the definition of ABM 'systems' and 'components' contained in Article II of the Treaty. (See the section on Article II, above.) The Reagan Administration now reads the definition as though the word 'currently' were simply not there: 'an ABM system is a system...consisting of...ABM interceptor missiles...ABM launchers...[and] ABM radars'. But in the Treaty's text, the phrase is 'currently consisting of', not 'consisting of'. In fact, as mentioned earlier, the word 'currently' was inserted precisely to make clear that the subsequent list is simply a description of the ABM components which were 'current' at the time, not a complete description of all possible components covered by the definition. The definition is clear. It covers all systems 'to counter strategic ballistic missiles or their elements in flight trajectory', regardless of the technology of their components. Moreover, agreed statement D, which is the only portion of the Treaty which makes an explicit reference to systems 'based on other physical principles', *explicitly refers to such exotic technologies as 'ABM systems' and 'components'*—that is, precisely what is limited by Article V and the rest of the Treaty.

The Reagan Administration bases its case on the negotiating record. Under the time-honoured rules of diplomatic privacy, this evidence is not available. In essence, the Administration argues that the Soviets were attempting to leave open a loophole so that they could later argue that exotic technologies were not covered by the Treaty. But if the

Soviets were intending to claim that exotic technologies were not ABM 'systems' and 'components', it would have been very sloppy negotiating indeed to allow the US to insert the word 'currently' into the definition of Article II, and to explicitly refer to future technologies as ABM 'systems' and 'components' in agreed statement D. In my experience the Soviets are not sloppy negotiators.

VIII. Exotic technologies under the traditional interpretation

The Treaty clearly bans the development, testing and deployment of all mobile ABM systems and components, regardless of their technology. Development and testing of all fixed, land-based ABM systems and components is allowed (except for missiles carrying multiple warheads or launchers capable of multiple fire or rapid reload), regardless of their technology.

In the case of ABM missiles, launchers and radars, the meaning of the word 'component' is reasonably clear. But when does a laser or particle beam become an ABM component, and therefore limited by the Treaty? Agreed statement D refers to future components as those things 'capable of substituting for' an ABM missile, launcher or radar. In the negotiations, it was agreed that technologies which were merely 'adjuncts', assisting ABM components without taking their place, were not limited by the Treaty. (The example used in the negotiations was a telescope used to assist a radar in discriminating warheads from decoys.) The concept 'capable of substituting for' provides a framework for thinking about the problem, but there are still vexing issues to be resolved. A laser might be capable of destroying a strategic missile (and thus substituting for an ABM interceptor) at a range of a few feet, but not at a hundred miles, which might be necessary for a practical ABM system. Should that laser be considered an ABM component? It may even be possible to divide the ABM task so that none of the pieces of a system handled all of the functions of an ABM interceptor, missile or radar. Many of the experiments planned for the next decade in the US Strategic Defense Initiative (SDI) programme raise precisely these issues. Resolving them will require clear thinking and political will on both sides—which may mean that they will go unresolved for some years to come.

IX. Article VI: non-ABM systems and components

During the ABM Treaty negotiations, the United States was concerned that ostensibly non-ABM systems could have significant ABM capabilities. The Soviets had deployed some 10 000 surface-to-air missile (SAM) launchers, supported by an extensive network of radars. Worst-case analysts in Washington had long worried that the Soviet SA-5 SAM system could be upgraded to have some ABM capability. The US could have had little confidence in limits on ABMs if the Soviets remained free to give anti-aircraft systems a capability to track and destroy re-entry vehicles (RVs) from US missiles. Soviet 'Hen House' early-warning radars were also judged to have some ABM capability. These concerns were eventually addressed in Article VI, in which each party undertakes:

(a) not to give missiles, launchers, or radars, other than ABM interceptor missiles, ABM launchers, or ABM radars, capabilities to counter strategic ballistic missiles or their elements in flight trajectory, and not to test them in an ABM mode; and

(b) not to deploy in the future radars for early warning of strategic ballistic missile attack except at locations along the periphery of its national territory and oriented outward.

Article VI was one of the most difficult and complex provisions of the Treaty to negotiate. We originally proposed strict numerical limitations on the capabilities of air-defence systems, veto power over deployments of all large phased-array radars, and intrusive inspection provisions. The Soviets resisted any constraints that might limit the capabilities of their air-defence network. The Treaty, they said, was to limit ABM systems, and there was no reason to limit the capabilities of non-ABM systems. Unlike the United States, which faces only a small and antiquated Soviet bomber force, the Soviet Union faces bomber threats from Europe and the Chinese, as well as large forces of US strategic and forward-based bombers. The Soviets have devoted enormous resources to the air-defence mission, and they were determined to prevent the Treaty from limiting the effectiveness of their air defence. Nevertheless, with persistence we were able to achieve the core of our objectives, a general prohibition on giving any non-ABM system an ABM capability, and on testing non-ABM systems in an ABM mode, as well as the geographical limit on deployment of early-warning radars.

It was clearly recognized, however, that Article VI did not entirely resolve the issue. One problem is that the phrase 'capabilities to counter strategic ballistic missiles or their elements in flight trajectory' is not defined. It is clear that Article VI does not forbid either side from developing systems to defend against short-range, non-strategic ballistic missiles. But the point at which such an anti-tactical ballistic missile (ATBM) system begins to acquire some capability against strategic missiles is not defined. Since intermediate-range missiles (IRBMs) such as the Soviet SS-4, SS-5 and SS-20 have ranges and trajectories similar to strategic submarine-launched missiles, a defence against such IRBMs would clearly be capable of defending against some submarine-launched ballistic missiles (SLBMs) as well, and would therefore be forbidden. But there is a broad grey area in which the ABM potential of ATBM systems is ambiguous. This area is becoming more significant as both sides press forward with ATBM programmes, the USSR with the SA-X-12, and the US with the upgrade of the Patriot air-defence missile. Both of these systems have now been tested against short-range ballistic missiles.

Similarly, the Treaty does not limit development, testing or deployment of anti-satellite (ASAT) weapons, as long as they are not given an ABM capability or tested in an ABM mode. But the task of detecting, tracking and destroying a satellite is not very different from that of shooting down a strategic missile RV in mid-flight. As a result, some types of ASATs would have an inherent ABM capability.

X. Testing 'in an ABM mode'

The phrase 'tested in an ABM mode' is used both in Article VI and the definition of ABM systems and components in Article II. It is central to the provisions of the Treaty. However, no agreed definition of this phrase was arrived at during the negotiations, although the United States made a unilateral statement (B) on the subject, giving examples of activities it would consider to be testing 'in an ABM mode'.

In the mid-1970s, US intelligence observed operations of the Soviet SA-5 air-defence radar at the Sary-Shagan test range which raised questions as to whether the radar was being tested 'in an ABM mode'. The United States raised the issue in the Standing Consultative Commission (SCC), a body created for such purposes by Article XIII. The activity in question immediately ceased, and an agreed statement defining 'tested in an ABM mode' was worked out. It was initialled in

1978, and is binding on both governments. (While the rules or the SCC require that the full text of the statement remain secret, the gist of the definition of 'tested in an ABM mode' can be found in the 1985 and 1986 Reports of the US Strategic Defense Initiative Organization, in the appendices discussing compliance with the ABM Treaty.) In 1985, a further agreement was signed which reportedly forbids *any* use of air-defence radars at test ranges when ABM testing is being conducted, except in the unlikely event that there are hostile aircraft in the area of the test range.

There is still some grey area, however. The point at which an ATBM system should be considered tested in an ABM mode is still somewhat ambiguous, and the 1978 agreed statement does not specify under what circumstances testing against satellites in space should be considered testing in an ABM mode. Several of the tests planned in the US Strategic Defense Initiative will intentionally exploit this ambiguity.

XI. Radar restraints

The Treaty's restraints on radars are of particular significance. Radars are the guiding eyes of present ABM systems. Since larger radars take years to build, their construction would provide adequate warning if one side was preparing to violate the Treaty. An agreement permitting only a small number of ABM launchers and interceptors would be less meaningful if a nation could build as many large radars as it wished. Widespread defences might be deployed rapidly, since additional interceptor missiles could be produced and stored covertly and then quickly deployed, with radars already available to guide them.

But large ABM-capable radars are also useful for a variety of other tasks, both military and civilian. This made agreed restraints more difficult to achieve. The Soviets complained that the radar limitations the US proposed would unduly constrain their air-defence systems. The US delegation argued that radars with ABM capabilities should be restrained, whatever their purpose. We said that ABM-capable radars could be identified by examining whether or not they were located so as to contribute to defence against missile attacks, whether or not they had a *phased-array type of antenna*, and whether or not they had a large *power-aperture product*, a technical parameter which determines the range at which the radar can detect incoming re-entry vehicles. The Treaty refers to the power-aperture products as 'potential', and defines it in agreed statement B as the product of the radar's mean emitted power in watts and the area of its antenna in square metres.

The final restraints on non-ABM radars were a compromise. Article VI covers early-warning radars, requiring that they be on the periphery of the country and oriented outward. Located that way, the radar's coverage would be almost entirely outside the country's territory, making it very difficult for it to serve effectively as a battle-manager for a missile defence. Moreover, early-warning radars on the periphery of a country's territory would be especially vulnerable to attack.

Agreed statement F broadens the restraints. It prohibits either country from deploying *any* new phased-array radars with a potential greater than three million watt-square metres, except as ABM radars under the terms of articles III and IV, early-warning radars under the terms of Article VI, or for space tracking and treaty verification. No phased-array radar for any other purpose may have a potential greater than three million watt-square metres. This greatly reduces the risk that either side could build up a radar base for a nation-wide ABM system from radars that were ostensibly deployed for non-ABM purposes. Unfortunately, however, no limits were placed on the location, number or capabilities of radars for space tracking and verification, and there is no agreed means of distinguishing between a space tracking or verification radar and an ABM or early-warning radar.

The Soviets are now violating the Treaty by building an early-warning radar near Krasnoyarsk, which is neither on the periphery of their country nor oriented outward. They have argued that the facility is intended for space tracking and is therefore allowed. But while it may have some capability to track objects in space, its design and orientation are clearly intended for early warning of ballistic missile attack. Therefore, even though the radar does not appear to be an ABM battle-management radar, and would not contribute significantly to providing a base for a nation-wide defence, it does constitute at least a technical violation of Article VI.

The US record in this regard is also deficient. At the time of writing, a large US phased-array early-warning radar is about to be completed at Thule, Greenland, and construction of another at Fylingdales Moor, United Kingdom, is about to begin. Needless to say, neither of these locations is on the periphery of the United States. The United States has argued that the new radars are merely modernizations of earlier radars that existed at those sites when the Treaty was signed, and hence are not 'future radars' as referred to in

Article VI. This argument is barely more plausible than the Soviet justification of the Krasnoyarsk radar. The new radars are very capable large phased-array radars—a type the Treaty clearly places in a special category—while the old radars are not. To treat them as mere modernizations seems questionable. In my view, they are new radars, prohibited by Article VI's ban on future early-warning radars at locations other than the national periphery and agreed statement F's prohibition on all new large phased-array radars not specifically permitted by the Treaty.

But like the Krasnoyarsk radar, these radars would not contribute significantly to the establishment of a nation-wide defence, even if they are determined to be technical violations of the Treaty. The issues raised by the radars on both sides are important and need to be resolved, but they do not go to the heart of the agreement. If political will and a modicum of ingenuity were applied to the task, I am confident that these questions could be satisfactorily settled, and the Treaty's crucial radar restraints reinforced.

XII. Articles VII and VIII: modernization, replacement and dismantlement

Article VII is simple enough: it allows the sides to modernize and replace their ABM systems and components, as long as the other provisions of the Treaty are met. Both sides are free to develop and deploy more sophisticated ABM radars, faster ABM interceptors, and so on. (In connection with the preceding discussion on early-warning radars, however, it should be noted that such radars are not included in the definition of the components of an ABM system, and hence their modernization is not authorized by Article VII.)

Article VIII commits each side to dismantling any excess ABM components beyond those allowed by the Treaty 'within the shortest possible agreed period of time'. Pursuant to this provision, the US ABM site at Malmstrom Air Force Base was dismantled soon after the Treaty entered into force. Negotiation of definitions, procedures and times for such dismantling is one of the key tasks assigned to the Standing Consultative Commission by Article XIII.

XIII. Article IX: non-transfer

Article IX commits each side not to deploy ABM systems or components 'outside its national territory', and not to transfer them to other states. This is amplified by agreed statement G, which extends the 'no-transfer' clause to cover the transfer of 'technical descriptions or blue prints specially worked out for the construction of ABM systems and their components'. This provision was included at the Soviets' behest. They had had considerable experience with the United States' deploying offensive weapons on the territory of its NATO allies, or transferring them outright to the United Kingdom. They did not wish to repeat the experience with ABMs. In an important unilateral statement, however, we made clear that this non-transfer provision should not be regarded as setting any precedent for limits on offensive arms.

This article too may be called into question by the US Strategic Defense Initiative. Several countries allied with the United States have recently agreed to participate in SDI research. Many of these arrangements involve 'teaming' of US and foreign companies in co-operative endeavours. Information is likely to be transferred in both directions. Once this work begins to move from the research stage to the development stage, it may begin to run up against agreed statement G's prohibitions on the transfer of 'technical descriptions' and 'blue prints' of ABM systems. Similar problems would arise if co-operatively-developed ATBM systems began to achieve some ABM capability.

XIV. Articles XII and XIII: verification and compliance

Article XII, covering verification of the Treaty, is a landmark in international arms control agreements. For the first time, the Soviets explicitly accepted the legitimacy of 'national technical means of verification', such as photoreconnaissance satellites or electronic intelligence systems—activities they had once considered espionage. Moreover, both sides agreed not to interfere with the other's national technical means, and not to use 'deliberate concealment measures which impede verification'. As a result of these provisions, the US can have greater confidence in its ability to detect Soviet weapon developments before they create a threat to US security. Both the SALT I Interim Agreement freezing offensive weapons and the SALT II Treaty contain similar provisions, but those agreements are no longer in effect. If the

ABM Treaty regime breaks down, one of the many tragic consequences will be an end to the agreed sanctuary and legitimacy granted to many US intelligence resources, which have made a major contribution to the stability and predictability of US Soviet relations.

No international agreement can foresee every contingency. Ambiguities and difficulties inevitably arise. For this reason, Article XIII created the Standing Consultative Commission with a broad mandate for discussion of issues related to the Treaty. (The SCC was given similar responsibilities under the terms of the SALT I Interim Agreement and SALT II.) It has served effectively in negotiating detailed provisions for dismantling of strategic forces covered by the agreements, negotiating agreed statements to remove ambiguities (such as the 'tested in an ABM mode' issue described earlier), and settling compliance disputes. By mutual agreement, proceedings of the SCC are secret, which has allowed problems to be discussed frankly, with a minimum of playing to the gallery.

In discussing compliance questions, the SCC is not and cannot be a court of law. It has no power to force sovereign nations to do anything they do not wish to do. But in the past, both parties have taken the Commission seriously, and a number of important compliance issues have been satisfactorily resolved. The Carter Administration issued a report in 1979 which concluded that in every case up to that time, when the United States raised questionable Soviet activities in the SCC, 'the activity ceased, or subsequent information clarified the situation and allayed our concern'. In recent years, however, the effectiveness of the SCC has declined. Too often the Reagan Administration has made public charges before all the facts were known and all avenues for possible settlement were explored. The emphasis has shifted from a search for mutually acceptable solutions to attempts to present a public brief against the Soviet Union and impose the solutions the United States would prefer. In dealing with a powerful Soviet Union, this is not likely to be a fruitful approach. If the SALT structure is to be maintained and extended, an institution such as the SCC will be vital. It should be nourished, not scorned.

XV. Articles X, XI and XVI: boilerplate

Articles X, XI and XVI are standard in US-Soviet agreements. Article X commits both sides not to undertake other obligations which conflict with the Treaty. Article XI calls for further negotiations on limiting offensive arms, signalling that SALT was not a one-time affair. Article XVI covers the details of ratification and entry into force.

XVI. Article XV: duration and withdrawal

As specified by Article XV, the ABM Treaty is of 'unlimited duration'. It is not simply a convenience of the moment, but a document based on conclusions that each side expected to be enduring. Either party may withdraw if 'extraordinary events related to the subject matter of the Treaty have jeopardized its supreme interests'. Withdrawal requires six months' notice and a statement of the 'extraordinary events' that are seen as justifying withdrawal. Under the accepted canons of international law, either party is also permitted to withdraw if the other commits a material breach of the agreement.

In 1972, it was felt that failure to limit the race in offensive arms might constitute an 'extraordinary event' that could jeopardize US supreme interests, and the US made a unilateral statement to that effect. Subsequently, the SALT II agreement was reached, which did provide a considerable brake on the offensive arms race. Nevertheless, today there are some who argue that the continued competition in offensive arms calls for withdrawal from the ABM Treaty. I cannot agree. As the competition in offensive arms continues, the arguments in favour of limitations in the defensive area only grow stronger. And without these limitations, we will surely be unable to restrain the competition in offensive strategic arms, or proceed to reductions in offensive strategic forces.

XVII. Conclusion

The ABM Treaty is a document of historic significance. It is a comprehensive, precisely drafted contract to govern ABM relations of the superpowers into the unlimited future. For as long as it endures, it rules out a race for defensive missile systems which had threatened to be a major new and dangerous form of arms competition. In the long reach of history, if the nuclear era lasts, it will seem like a surprisingly sensible thing that the superpowers did in 1972—agreeing not to duplicate in the defensive field the foolish, costly, dangerous escalating competition that they had been slaves to for over 20 years in the offensive weapons field.

Even more important, the Treaty is the corner-stone for all efforts to control and reduce offensive strategic arms. No military planner is likely to look favourably on proposals to limit or reduce his offensive forces when the adversary's defences are unrestrained. It is futile to expect reductions in offensive arms, or even the maintenance of existing offensive agreements, unless the ABM Treaty is maintained.

Tragically, the survival of the Treaty is very much in doubt. The current US Administration has embarked on a massive programme of ABM research and development under the aegis of the Strategic Defense Initiative. Senior officials have expressed firm expectations that this will lead to deployment of defences and an end to the Treaty. The Treaty has been unilaterally reinterpreted in a way that would gut its key provisions. At Reykjavik, the Administration demanded that the Soviets accept a major revision of the ABM Treaty, creating a fundamentally different document that would allow unlimited development and testing of space-based exotic-technology ABMs, and would terminate the Treaty after 10 years, leaving each side free to deploy whatever defences it pleased. The Reagan Administration has clearly set itself on a course aimed at the demolition of the ABM Treaty.

To continue on this course would be a blunder of catastrophic proportions. After Reykjavik, the alternatives are stark: we can either have arms control or we can have a crash programme to deploy defences. We cannot have both. The end of the ABM Treaty would mean the renewal of a costly and dangerous arms race in both offences and defences, with no agreed restraints of any kind. One can only hope that over the coming months the two sides will come back once again to the common ground found in 1972, and agree to maintain and strengthen the ABM Treaty.

Notes and references

[1]   Ambassador Smith was the US negotiator of the 1972 ABM Treaty.

Paper 3. The Treaty's basic provisions: view of the Soviet negotiator

Vladimir Semenov[1]
Ministry of Foreign Affairs, Moscow, USSR

I. Achievement of strategic parity

By the beginning of the 1970s a new military-political situation had been created throughout the world, in which the most important stabilizing factor was the achievement by the Soviet Union of strategic parity with the USA. This made possible the first negotiations for the limiting of strategic weapons (SALT I).

II. Recognition of the offensive-defensive link

As a result of many years' discussions between Soviet and US leaders and their delegations, both parties reached an understanding that with parity of strategic offensive weapons deployment by one of the parties of additional defence potential would be tantamount to gaining the capability for a preventive first strike. Logically, nuclear confrontation implies that a close link exists between offensive and defensive strategic systems.

Recognition of this fact led to mutual understanding of the necessity to concentrate efforts on preparing an agreement to strictly limit anti-missile defence systems. The parties agreed that the creation of a ramified anti-missile defence system would lead to an undermining of the strategic balance and would destabilize the global strategic situation. There would be no doubt that the other side would then be forced to strengthen its strategic offensive potential and increase its means of counteraction against anti-missile defence systems in order to restore the parity that had been broken. All this would lead to an unlimited arms race. In this way, an agreement to limit anti-missile defence systems was an objective necessity of a nuclear missile epoch. This mutual understanding resulted in the signing on 26 May 1972 of an agreement on the limitation of anti-missile defence systems, which was of indefinite duration, and an interim agreement on certain measures to limit strategic offensive weapons (SALT I).

The agreement about anti-missile defence became the corner-stone of the whole process of limiting and reducing nuclear weapons. By signing this agreement, the Soviet Union and the USA confirmed unanimously that in the century of nuclear weapons only mutual restraint within the area of anti-missile defence systems gives the possibility of moving forward along the road towards limiting and reducing nuclear weapons.

The objective interdependence between offensive and defensive strategic systems was reflected not only in the concrete provisions of the SALT agreement but also in other Soviet—US agreements. In a mutual Soviet—US communiqué of 30 May 1972, the United States and the Soviet Union confirmed that the agreement limiting anti-missile defence systems and an interim agreement on some measures limiting strategic offensive weapons represented a big step in the direction of restraining and eventually halting the arms race. They correspond to vitally important interests of the American and Soviet people and the people of other nations. SALT II was also based on an anti-missile defence agreement.

III. The Treaty—legal viewpoint

Let us proceed to the actual anti-missile defence agreement text and examine it from a legal point of view.

First and foremost, one would wish to give attention to the following: during discussions about the meaning and role of the agreement to restrain anti-missile defence systems, one often neglects to mention the fact that it is an international juridical document of indefinite duration. Both parties then emphasized their mutual point of view about this fundamental, permanent meaning and key role in the whole process of restraining strategic weapons. Doubts are being expressed about the 'further fate' of the anti-missile defence agreement. These are clearly calculated to undermine this agreement. They are an attempt to depart from previously undertaken obligations and cover plans and actions incompatible with the agreement.

In the preamble to the Treaty, a concept now denied by enemies of nuclear disarmament is clearly expressed: 'effective measures to limit anti-ballistic missile systems would be a substantial factor in curbing the race in strategic offensive arms' and would reduce the danger of a nuclear war arising.

It is already 15 years since the anti-missile defence agreement was made and time has proved this conclusion to be correct. During the whole period, the agreement has been a pivot in strategic stability and international security, a foundation in the process of restraining nuclear weapons, and a basis for strategic relations between the Soviet Union and the United States. The above-mentioned quotation from the preamble was a recognition of the indissolubility of the mutual links between strategic offensive and defensive weapons and of the role of anti-missile defensive systems as a catalyst to the arms race. Statements that this provision was only a result of the absence at that time of technical possibilities to create effective anti-missile defence systems have no basis. The linkage between offensive and defensive systems has a fundamental and permanent character and would be of even more immediate interest with the appearance of possibilities to create more technically complicated anti-missile defence systems. If one side created anti-missile defence systems with capabilities above the level permitted by the Anti-Ballistic Missile (ABM) Treaty this would inevitably lead to the destruction of the strategic balance and an increased risk of nuclear war. This key idea of the anti-missile defence agreement suited the conditions of 1972 and is even more appropriate today.

Briefly, the main *limitations* imposed by the agreement are mentioned in articles I and V. According to Article I, the Soviet Union and the United States undertook an obligation of unlimited duration to limit anti-missile defence systems, not to deploy anti-missile defence systems on their own territory and not to create a basis for such defence. This general obligation entails that each party shall not deploy anti-missile defences for areas other than those stipulated in Article III, that is, two areas were excepted: an area centred on the national capital, and a deployment area containing intercontinental ballistic missile silo launchers. These areas were limited by size, and limitations were stipulated for components of anti-missile defence systems positioned in each area. Later on, with a desire to contribute to the goal of the agreement, progress during the negotiations on limiting strategic offensive weapons and consolidation of international peace and security, the parties agreed to limit deployment of permitted anti-missile defence systems to only one area, not two areas for each side as stipulated in Article III of the agreement. This commitment was stated in the protocol of the anti-missile defence agreement of 3 July 1974. The protocol came into force on 25 May 1976 and is regarded as an integral part of the agreement. In Article I of the protocol, the Soviet Union and the USA committed themselves not to exercise their right to deploy an anti-missile defence system or its components in the second of two areas where the positioning of anti-missile defence systems was permitted by Article III of the Treaty, except in the case of substitution of one of the permitted areas for the other. As we see, in the strengthening of the dynamics of the parties' obligations, they aspired maximally to a strict limitation of an anti-missile defence system. Claims that the Strategic Defense Initiative (SDI), which has a goal of creating a large-scale anti-missile defence system with elements based in space, does not contradict these obligations cannot be taken seriously from a legal point of view. The SDI programme, by emphasizing the creation of a large-scale anti-missile defence system with elements based in space, directly contradicts the clear general prohibition according to Article I.

The 'new interpretations' of the clause in Article V, asserting that it concerns only the components of an anti-missile defence system which the parties had in their defence arsenals in 1972, are unscrupulous distortions. To substantiate this, let us investigate this so-called new interpretation. For example, it is alleged that the clause in Article V is only relevant for traditional anti-missile defence components named in Article II of the agreement, that is for anti-missile missiles, their launchers and anti-missile system radar stations.

In reality, the above-mentioned enumeration means that 'currently' (on the date the agreement was signed) an anti-missile defence consisted of these components. The word 'currently', according to the negotiation record, was included in the text precisely to avoid the inclusion of the enumerated specific components in a general definition of the ABM defence system and to thereby avoid freezing the definition at the 1972 level of knowledge—the anti-missile defence agreement is, after all, a document of indefinite duration.

The general definition of anti-missile defence was given in Article II: 'For the purposes of this Treaty an ABM system is a system to counter strategic ballistic missiles or their elements in flight trajectory...'. We can conclude that this definition of an anti-missile defence system excludes any possibility of interpreting the provisions of Article V as referring only to those components of anti-missile defence which were in the parties' weapon inventories in 1972.

One would wish to show with the help of the agreements' provisions and negotiating history, the great effectiveness and vitality of the anti-missile defence agreement. The authors wisely foresaw the possibility, procedure and place for discussions, specifications and solution of questions concerning fulfilment of commitments.

Article II of the agreement includes a general definition of an anti-missile defence system and names their components—ABM interceptor missiles, launchers for ABM interceptor missiles and anti-missile defence radar stations. When the necessity to specify the term 'tested in an ABM mode' arose, the parties, in a constructive and business-like spirit, worked out and signed such specifications in the agreed statement of 1 November 1978.

One more fact: Article IV mentions anti-missile defence systems or their components used for development and testing. During the negotiations, questions arose concerning the definition of test ranges and the possibility of placing on the test ranges other types of weapons or military technology for specified testing or to guarantee the security of the test range.

Questions also arose in connection with Article VI, which includes a commitment not to give to these components of an anti-missile defence system the capability to counter strategic ballistic missiles or their elements in flight trajectory, and not to test them for anti-missile defence purposes. It also restricts deployment of radar stations which warn of missile attack. In these cases too, the parties could, as a result of careful discussions in a permanent Soviet-US consultative committee, work out within the anti-missile defence agreement mutual understanding in questions arising in connection with articles II, IV and VI. This understanding was fixed in the parties' agreed statement of 1 November 1978. Incidentally, one can say that fairly recently (1985), the parties in the permanent consultative committee devised the general understanding of 6 July 1985 concerning the question of incompatibility of anti-missile defence system components and air-defence system components working simultaneously on the testing ground. From these facts, one can see that when political will exists the agreement permits solution of every question arising, thus demonstrating its enormous potential, viability and effectiveness.

When analysing the Treaty text and its place in the present international situation, one has also to mention articles IX and X. To guarantee the agreement's viability and effectiveness, the parties committed themselves not to give or transfer to other countries and not place outside their own territory anti-missile defence systems or their components, and not to undertake international commitments which would contradict the agreement. In the light of those commitments, how should one judge the US action of involving other countries in work on the SDI programme? It can be seen in one way only—as directly contradicting those commitments. Such action by the USA makes other countries participants in the violation of a document of most vital global significance.

In international law there is no doubt that legal commitments in accordance with the agreement have a dominating role and apply to additional documents. The components of the SALT I agreements are not only the anti-missile defence agreement and the interim agreement, but also, in particular, the parties' agreed statements supplementary to those. One of these statements (commitment 'D') is being used by opponents to the anti-missile defence agreement in the USA as a justification for the SDI programme.

Agreed statement D states that to guarantee fulfilment of the obligation not to deploy anti-missile defence systems and their components, except as stipulated in Article III of the Treaty, the parties agree that in the case of creating in the future anti-missile defence systems based on other physical principles which were capable of replacing ABM missiles, launchers for ABM missiles and anti-missile defence radar stations, concrete limitation of systems and their components would be subject to discussion in accordance with Article XIII and agreement in accordance with Article XIV of the Treaty.

In 1985, the US Administration declared the existence of 'a new interpretation' of the anti-missile defence agreement which had been prepared by the State Department's legal adviser A. Sapphire. In this interpretation statement D is interpreted as supposedly permitting unlimited creation and almost deployment of anti-missile defence systems based on other physical principles. We shall also discuss this fiction.

In the so-called new interpretation emphasis is put only on that part of the statement that says that in the case of the creation in the future of anti-missile defence systems based on different physical principles the parties will discuss the concrete limitation of such systems and their components, deliberately ignoring the initial part of the statement which says 'In order to insure fulfilment of the obligation not to deploy ABM systems and their components *except as provided in Article III of the Treaty...*' (author's emphasis).

From the initial part, the unambiguous conclusion can be drawn that the rest of the statement concerns those anti-missile defence systems and components 'based on other physical principles' which are allowed in Article III, that is only those within one area around the national capital or one area for the deployment of intercontinental ballistic missile silo launchers. Space-based systems or components cannot be deployed in the areas stipulated in Article III. Consequently it is illegal to apply provisions of agreed statement D to space.

As head of the Soviet delegation for the SALT I negotiations I can see with full responsibility that the new 'interpretation' is directly contrary to the initial meaning which the parties expressed in the statement and that it does not correspond to the documents of the negotiations.

In the above-mentioned attempts of the present US Administration a pattern of change in the meaning of the agreement can be clearly seen—the main content of the agreement is being destroyed.

As we know, the Soviet Union always opposes every action which would lead to undermining the anti-missile defence agreement and is against making the agreement a shield for the US policy directed at an arms race on earth and in space.

In fact, the participants of the negotiations, when agreeing on statement D, looked into the future and strove to strengthen obligations in accordance with the agreement. They did not exclude the possible appearance in the future of anti-missile weapons based on other physical principles. But those permitted would be connected to anti-missile defence only in an area permitted by the agreement and they would be fixed ground systems, not space or other systems.

In this way, the intention of statement D was and is to increase and reinforce the clause of the agreement that forbids deployment of any large-scale anti-missile defence system and was not meant to abolish the key prohibition of Articles I and V. The opposite approach lacks all legal and logical sense. It is generally accepted that in international legal documents one does not include contradictory commitments. If the reverse were true, no-one would devise such a document.

Analysis of the fundamental clauses of the anti-missile device agreement totally disproves claims that the SDI programme does not contradict the anti-missile defence agreement. This programme contradicts both the spirit of the agreement and its key clauses, as discussed below.

Firstly, the declared goal of SDI is the deployment of an anti-missile defence system for the country's territory. *Article I(2)* of the agreement directly forbids deployment of such a system or creation of a base for such a defence.

Secondly, the SDI programme anticipates the creation of an anti-missile defence system with space-based elements. Article V(1) of the agreement forbids creation, testing and deployment of an anti-missile defence system or components which are sea-, air-, space- or mobile-land based.

Thirdly, different 'interpretations' to justify creation of anti-missile defence space components are illegal. All clauses limiting anti-missile defence systems constitute a single whole and are intended to guarantee fulfilment of the main commitment in accordance with Article I.

Agreed statement D is an integral part of the anti-missile defence agreement and cannot contradict its provisions, in particular also *Article III*. This means that the agreement states that anti-missile defence systems and components based on other physical principles are applicable only to fixed ground variants and to one permitted anti-missile defence area.

The attempts to present the SDI programme as harmless 'scientific research' are groundless. The character of US work already being carried out within the SDI programme testifies clearly that the process of creating space components in a large-scale anti-missile defence system is being speedily realized; that which is being done is directly forbidden by the anti-missile defence agreement.

IV. Implications for international security

The consolidation of international security and strategic security depends to a considerable extent on maintaining an agreed base for the limitation of arms—consolidation of that which already exists in this area. Of decisive importance is the maintenance and consolidation of the agreement limiting anti-missile defence systems.

The importance of the Treaty is enormous. This is the corner-stone of the whole process of limiting and reducing strategic weapons. By signing the agreement, the Soviet Union and the United States demonstrated their conviction that in this nuclear century only mutual restraint in the anti-missile defence area will give the possibility of moving forward on the road towards reduction of nuclear weapons.

If the anti-missile defence agreement loses its force, then the basis for the reduction of nuclear weapons will disappear, which would mean an uncontrolled arms race. In the light of this, the question of reinforcing the anti-missile defence agreement regime achieves great importance. How do the parties react to this necessity?

The Soviet side has proposed reinforcing the ABM Treaty regime through acceptance by the parties of the obligation not to use the right to retire from the agreement for at least 10 years, with rigorous observance of all the clauses in this document. This means the observance of the prohibition on creating, testing and deploying anti-missile defence systems and components based in space. In other words, we have suggested that all the work within the SDI programme should be limited to laboratory research.

By putting forward at the meeting in Reykjavik an integral package of suggestions, the key element of which was the demand to reinforce the ABM Treaty regime, the Soviet side had in mind the necessity of creating a situation in which there would not be attempts to undermine strategic stability and upset agreements, including agreements in the area of strategic offensive weapons and medium-range missiles. For this purpose, testing in space of all anti-missile defence space elements would be prohibited, apart from research and testing in laboratories; the Soviet Union and the United States would relinquish the right to withdraw from the anti-missile defence agreement for a 10-year period, in the event of which their strategic offensive weapons would be dismantled in two stages. After this period, during special negotiations, mutually acceptable decisions concerning the next step would be worked out.

However, because of the US standpoint, agreement on the matter was not achieved at Reykjavik—the Americans did not agree to limiting research, development and testing within the SDI programme to the laboratory.

The US standpoint concerning space weapons reflects an attempt to use all means to leave open the channels for putting weapons in space. In this connection, the USA treated negatively our propositions about prohibiting space strike weapons as a whole, and about reaching agreements prohibiting space weapons of the type 'space-earth' and anti-satellite systems.

Verbally, the US side supports the prevention of the arms race, but in reality it simplifies this problem by suggesting agreement on 'regulated' deployment of space weapons.

The USA denies the objective linkage between defensive and offensive weapons and attempts to present US strategic 'defences' as some independently acting factor, which has, moreover, some special 'stabilizing' functions. By doing this, they side-step the fact that 'defensive means' based on modern technology may have a significant offensive potential (take, for example, space weapons of the class 'space-earth').

Since the meeting in Reykjavik it has been stated that the USA appears to support preservation of the anti-missile defence agreement in an unchanged form and observance of all the Treaty provisions during a specified period. Outwardly, one gets the impression that the parties stand close to one another, but the main difference is the interpretation the parties put on the purpose of not abandoning the agreement for 10 years.

In suggesting the obligation not to leave the agreement, we assume that strategic offensive weapons will be abolished during a 10-year period. Herewith, the boundary between the permitted and the forbidden work has been drawn so as to limit work to the laboratory. Creating and testing of anti-missile defence systems and components with space bases outside the laboratory have to be prohibited because they do not correspond to the limitations of the anti-missile defence agreement.

The US is of another opinion. In accordance with this, research, development and testing of 'more sophisticated means of strategic defence' may be carried out both in laboratories and outside. This is an attempt to legalize every kind of work within the SDI programme. And it is the quintessence of the difference of opinions. This is a matter of principle.

In its approach to reinforcing the ABM Treaty regime, the consistent line of the Soviet Union on nuclear disarmament has been prominent. The direct road to the achievement of these goals is through upholding the agreement and refraining from the intention of transferring the arms race into space.

Unfortunately, a negative picture of the US approach has recently become evident. Now, for example, discussion is taking place in the USA concerning the Pentagon proposition of partial deployment of anti-missile defence systems with space-based elements already in the early 1990s.

Such a decision would not only be a step backwards from the agreements achieved in Reykjavik, but a complete about-turn. Such a situation would eliminate the possibility of reaching agreement concerning the reduction of nuclear weapons. What is more, it would eliminate the basis for discussion between the United States and the Soviet Union concerning nuclear weapons in space. In fact, US deployment of anti-missile defences in space would negate one of the key elements of the Geneva negotiations—the prevention of an arms race in space—and would render them pointless.

Quite recently the US side, as its main argument for advancing SDI, referred to its intention of replacing nuclear weapons as a deterrent with non-nuclear defence forces. This argument has now been totally discarded. Supporters in the USA of gradual deployment of anti-missile defence systems are already saying that this system has to strengthen US nuclear potential and that nuclear offensive weapons will be retained and supplemented with defensive weapons. The camouflage of arguments and promises was thrown away. The real face of SDI appeared before mankind. Comments are not necessary.

One hopes, however, that the voice of common sense will prevail in the United States. The whole world would benefit if the USA, not only in words but in actions, were to show its fidelity to the provisions of the anti-missile defence agreement. At the present stage of historical development, the importance of an anti-missile defence agreement to the destiny of the world is so great that it must be considered not only as a bilateral agreement but as the mutual property of the whole of mankind. This is what it has been, in fact, during its 15-year existence.

Notes and references

1   Ambassador Semenov was the Soviet negotiator of the 1972 Treaty. This paper is a translation of the Russian text.

Part III. Current US and Soviet ballistic missile defence programmes

Paper 4. Soviet research and development of directed-energy weapons

Simon Kassel[1]

Rand Corporation, 2100 M. Street NW, Washington, DC, 20037, USA

I. Introduction

The purpose of this paper is to consider the question of Soviet development of directed-energy weapons, based on information available in Soviet technical literature. Because the latter does not refer to such weapons in any form, one must survey the broad scope of pertinent Soviet research and development efforts and the existing technology that provide the necessary foundation of the several directed-energy weapon concepts known today.

The interest in directed-energy weapons is due to two reasons. First, these weapons represent totally novel concepts embedded in an advanced ballistic missile defence (BMD) programme, such as the Strategic Defense Initiative (SDI), and contribute significantly to the distinction between the latter and the missile-based BMD systems. Other components of advanced BMD proposals, comprising traditional interceptor missiles, curtains of projectiles or drifting debris barring the path of attacking warheads, or small homing rocket vehicles on mid-course interception missions, have roots in the older BMD systems and have been the subject of many analyses in the past.[2] Second, the Soviet Union has invested considerable effort in the development of directed-energy technology, which is amply reflected in Soviet technical literature. Thus, if some of that development has indeed been intended for the weapons application, it would represent the closest Soviet approach to the advanced SDI ideas that can be gleaned from the available information.

The concept of directed-energy weapons is still fraught with considerable uncertainty relating to its technical feasibility, practicality and cost. The degree of uncertainty varies with the type of weapon and increases with the successive stages of development. It is therefore important, when discussing the status of directed-energy weapons development, to make clear distinctions among the developmental stages as well as among the weapon types. One could roughly consider three such stages: the basic device projecting an energy beam with the required kill capability; the prototype weapon meeting the field requirements of primary power, weight and size; and the weapon system incorporating target aquisition, pointing and tracking capabilities. Since the uncertainties involved and the limitations of Soviet literature largely exclude consideration of the second and third stages, the following review deals mainly with the first.

II. The context of Soviet directed-energy development

Soviet directed-energy technology was born in the 1960s in a political climate that was openly favourable to the idea of creating a comprehensive and effective ballistic missile defence, regardless of its cost. The attitude of Soviet leadership was illustrated by Kosygin's 1967 remark that 'Ballistic missile defense may be more expensive than an offensive system, yet it is intended not to kill people but to save lives'.[3]

One impassioned Soviet plea of the time is strongly reminiscent of today's arguments in favour of SDI: 'It is theoretically and technically quite possible to counterbalance the absolute weapons of attack with equally absolute weapons of defense…The creation of an effective anti-missile system enables the state to make its defenses dependent chiefly on its own possibilities, and not only on mutual deterrence…'.[4]

Soviet military writers of the time also openly addressed the concept of directed-energy weapons: 'Powerful ground radar stations can produce plasma that will arise around a ballistic missile…Under the effect of the energy produced by the plasma, the ballistic missile will either be destroyed or knocked off the flight trajectory'.[5] And: 'If a method of focusing large amounts of energy over considerable distances is developed, it will be possible to resolve many scientific and technical questions, and especially the problem of destroying intercontinental missiles'.[6]

Along with the favourable political and military attitudes prevalent in the mid-1960s, Soviet science has also provided a nurturing environment for the concepts of directed energy. The principal promotor and developer of directed-energy technology in the USSR has been the Soviet Academy of Sciences and its network of research institutes. In the world of science, the Academy is a unique organization, being both a co-ordinator of national R&D and a performer of leading-edge research.

In the latter capacity, it has been responsible for the development of practically all the advanced technologies in the Soviet Union. As an autonomous scientific establishment, the Academy has been independent of the industry with its centralized control and ever-present production quota constraints, and therefore more able and willing to pursue high-risk research ventures. The Academy's approach to R&D also differs considerably from the style of the traditional Soviet military R&D establishment, characterized by conservatism and incremental improvements of existing design. The Academy science has been bolder in seizing new ideas and transforming them early into experimental prototype systems. These are the necessary preconditions for the development of directed-energy weapons.

The practical realization of all the known variants of directed-energy weapons depends on what is known as pulsed-power technology, the array of principles, techniques, and equipment necessary to generate and shape electrical pulses of megawatt-to-terawatt power that are the foundation of high-current electron accelerators and other directed-energy devices. This generic technology is discussed here first, followed by a review of the three main weapon aspects of directed energy: charged particle beams, neutral particle beams, and electromagnetic energy beams which include laser and microwave frequency domains.

III. Pulsed-power technology

The Soviet Academy of Sciences has been developing pulsed-power technology since the late 1950s. Within a decade, Soviet leadership accorded this effect an unprecedented recognition by including it as a major component in the overall electric energy development plans. The national energy R&D programme has been administered by the Department of Physico-technical Energy Problems of the Soviet Academy of Sciences. In the area of electric energy, the Scientific Council on Theoretical and Electro-Physical Problems of Electric Power has been the principal executive arm of the Department. Approximately one-third of the Council's range of interests has been dedicated to pulsed power.[7] The Council's chairman was M.P. Kostenko, who together with Petr Kapitsa designed the first practical pulsed-power generator in the Soviet Union. The generator produced a pulse shape necessary for the introduction of high-intensity magnetic fields. Kostenko's 1968 appointment to head the Council reflected an emphasis on pulsed power from the Council's inception.[8] In addition to his chairmanship of the Council,

Kostenko was also chairman of the Council's Working Section on High-Voltage Electrophysics. The Section's goals for the period of 1975-1980 were the development of megavolt nanosecond-pulse generators, capacitive and inductive energy storage, generation of super-strong magnetic fields, and the study of high-field behaviour of metals and dielectrics. All these topics lie at the core of the fundamental engineering problems of directed-energy weapons and their effects.

A major figure in Soviet pulsed-power and directed-energy field is Ye. P. Velikhov, vice-president of the Soviet Academy of Sciences, and board member of the State Committee for Science and Technology. Velikhov has been one of the leaders of the inertial fusion programme and the author of scientific papers on pulsed-power applications that may be relevant to directed-energy weapons.

Soviet pulsed-power R&D has involved over 30 institutes of the Academy of Sciences aided by research institutions of the universities, the State Committee for Atomic Energy, and a few industrial organizations. In Moscow, the leaders are the Lebedev Physics Institute, the birthplace of Soviet laser research and a major centre of work on high-current electron beam dynamics under M. S. Rabinovich and A. A. Rukhadze, the Kurchatov Institute of Atomic Energy with a large electron-beam programme under Velikhov and L.I. Rudakov, the Engineering Physics Institute, and others. In Leningrad, the Yefremov Institute of Electrophysical Equipment builds large accelerators. In Khar'kov, the Physico-technical Institute of the Ukrainian Academy of Sciences performs innovative work on accelerators and beam behaviour studies under Ya. B. Faynberg and Yu. V. Tkach.

Significant pulsed-power work is performed in Siberia. Novosibirsk is the home of the Nuclear Physics Institute which under the leadership of the late G.I. Budker and now under A.N. Skrinskiy has been responsible for many pioneering accelerator ideas and high-current switch designs. Another member of the Novosibirsk team is the Institute of Hydrodynamics, which is developing magnetic flux compression generators, useful in supplying high-current pulses. Tomsk represents a large concentration of pulsed-power development facilities, including the Institute of High-Current Electronics under G. A. Mesyats, the only facility entirely dedicated to pulsed-power and its applications. In the mid-1970s, the total manpower in the USSR engaged in the development of pulsed-power systems may well have approached 2000 scientists and engineers.[9]

Significant work in this field has also been performed by many other organizations of the Academy of Sciences, the research institutes of the university system and some industrial facilities. Together, the Soviet research institutions have addressed all the key components of pulsed power required for the known applications of directed energy. Some of the highlights of this research that are relevant to directed-energy weapons can be summarized as follows.

Primary power sources

While early experimental systems are stationary and can draw their energy from commercial grids, many pulsed-power applications are intended to be transportable and require efficient, autonomous primary power sources. A unique concept of such a source developed in the USSR is the pulsed magnetohydrodynamic (MHD) generator consisting of a rocket motor whose nozzle is immersed in a magnetic field and fitted with electrodes. The conductive exhaust gas of the motor cuts the field lines and generates electricity. The velocity and volume of the exhaust gas can translate into very high output powers.

Progressively refined versions of such a generator, initially called the Pamir series, were developed by Velikhov in the early 1970s at the Kurchatov Institute.[10] According to Velikhov, the Pamir-1 generator delivered electrical pulses of 100 megajoules and about 10 megawatts. These were to be injected into the earth's crust to study geological formations and seismic phenomena. Recent versions of the generator are said to deliver 15 MW.[11]

Energy storage

The conventional energy store of pulsed-power systems, represented by a large bank of capacitors, carries a large size and weight penalty. Soviet scientists have pioneered inductors as an alternative form of energy storage. Inductive storage systems are superior to capacitive storage ones because their energy density is much higher than that of the capacitor bank.

Switching

The performance of all pulsed-power applications, including high-current particle accelerators, high-energy lasers and macroparticle railguns, critically depends on the design of opening and closing switches to ensure the precise timing, shape, frequency and size of the output bolt of energy. Because of the stringent performance requirements, switch design still challenges the developers of directed-energy systems. In the 10 years after 1965, over 40 per cent of all Soviet authors who published papers in the pulsed-power field wrote on switching. Soviet switch designs, particularly those intended for application to inductive storage systems, were characterized by extensive variety and originality of design solutions.

Many pulsed-power components required for directed-energy weapons are also needed, in some form, for controlled fusion experiments. Soviet literature often cites the latter as the rationale for developing these components. Nevertheless, there are many aspects of Soviet pulsed-power research that indirectly indicate a weapons application. The apparent drive to reduce the size and weight of pulsed-power equipment while pressing for ever higher output energies and the variety and scope of Soviet research efforts go well beyond the requirements of controlled fusion. An early indication was the work on repetitive high-power switches. Neither small size and weight nor repetitive switches were relevant to the early stage of inertial fusion research, yet both were essential for the development of directed-energy weapons. While Velikhov's MHD generator has been developed for geophysical research and its qualifications as the optimal prime power source for directed-energy devices are not clear, its high output power and portability are of potential interest to such applications.

Soviet preference for inductive storage as a solution to the power supply problem also indicates a possible weapons orientation of their research, partly because of size and weight considerations. Inductive storage, in turn, demands an intensive development of advanced switching, the abundant evidence of which in Soviet R&D may thus be interpreted as willingness on the part of the Soviets to pay the price to minimize the size and weight of their pulsed-power systems.[12]

IV. Electron beams

The high-current electron-beam accelerator, as distinct from low-current high-voltage accelerators known in basic physics research, is a major product of pulsed-power technology. It is the source of electron beams intended to propagate in the atmosphere, in one variant of directed-energy weapons, and to pump free-electron lasers and drive high-power microwave devices, in other variants. The accelerator has also been from its inception the workhorse of experimental studies of high-current charged particle beam generation and behaviour. The earliest large Soviet electron accelerator of this type appeared in the literature in 1971; this was RIUS-5, rated at 4 MeV and 30 kA, and developed by Ye. A. Abramyan at the Nuclear Physics Institute in Novosibirsk. Six years later, the last such machine to be reported, the LIU-10, was put in operation. The LIU-10 was a radial-line induction accelerator, consisting of 14 induction modules delivering a total beam energy of 13.5 MeV at 50 kA.[13] Besides being nearly six times more powerful than RIUS-5, this terawatt machine represented a pioneering design and a significant step forward in the state of the art at the time. Work on the LIU-10 must have commenced in the 1960s; the patent application for it was submitted in 1968.[14]

The Soviet effort to design high-current pulsed electron accelerators has been in evidence since the mid-1960s and showed a steadily increasing growth rate up to the mid-1970s, at least from the viewpoint of publication frequency. After that time, Soviet publications on high-current electron accelerators steadily declined. These machines were initially developed for the inertial fusion programme which postulated a beam of electrons compressing and heating a pellet of nuclear fuel to initiate thermonuclear reaction. The programme has been led by Rudakov at the Kurchatov Institute of Atomic Energy.[15] Experience gained in the course of this work soon generated interest in other areas of application where electron beams could revolutionize the conventional methods and significantly advance the current performance levels. Such is the effort led by Rukhadze at the Lebedev Physics Institute, which offers the promise of generating very high microwave power from the interaction of electron beams and plasma. The Soviets expect eventually to generate by this means 100 gigawatt microwave pulses with a beam-to-microwave conversion efficiency of 10 per cent.[16]

Another such effort is acceleration of ions by electron beams, based on the collective acceleration principle invented by V. I. Veksler, the Soviet pioneer of synchrotron machines for high-energy physics experiments. Collective acceleration is expected to significantly increase beam energy, decrease the cost of particle accelerators, and, most importantly, reduce the size of the accelerators by increasing the voltage gradient. Soviet plans are to reach voltage gradients as high as 100

MV/m. Very high beam energies and, especially, high voltage gradients are essential for particle beam weapons.

Using high-current electron accelerators, Soviet physicists pioneered the production of megabar pulse pressures in solids and performed systematic studies of beam-materials interaction, all necessary to understand the mechanism of target destruction by electron beams.

An interesting Soviet research programme concerned ionospheric sounding with electron beams and generation of artificial auroras. An experiment with the code name Zarnitsa (Summer Lightning) was performed in 1973 with an electron accelerator aboard an MR-12 meteorological rocket. At an altitude of 100 km, the accelerator injected a 9 kV, 0.5 A pulse along the geomagnetic lines.[17]

Parallel with the application of electron beam accelerators, the Soviets have continued intensive theoretical and experimental studies of high-current electron beam behaviour in vacuum, in a variety of gas species, and most notably, in air, the latter of critical importance only to the problem of propagating the endo-atmospheric weapon beam.

The early Soviet development of high-current electron accelerators culminating in the LIU-10 would lead one to expect new generations of this machine and continuing effort. However, references to new high-current accelerators of the LIU-10 class disappeared from Soviet literature in the 1980s.

V. Neutral particle beams

High-current charged particle beams are not suitable for propagation in space. The environment of space, devoid of an appreciable atmosphere but containing geomagnetic field, requires for this purpose a beam of neutral particles. To acquire the necessary velocity and directional control, the neutral particles must first be generated and accelerated as charged particles, usually negative ions, which are then stripped of their extra electrons. The sequence of generation, acceleration and stripping is the basis of the neutral particle weapon beam; in each of these steps the Soviets have made a significant contribution to the state of the art. Particularly notable is their invention of the high-brightness, surface-plasma ion source and the radio-frequency quadrupole (RFQ) ion accelerator system.

In 1971 V.G. Dudnikov of the Nuclear Physics Institute in Novosibirsk obtained intense negative ion yields by introducing caesium vapour into a specially designed arc discharge source, which he called a surface-plasma ion source. This was one of the earliest known devices with a sufficient potential for consideration in the neutral particle beam weapon.[18]

The radio-frequency quadrupole accelerator suppresses the divergence of the particle beam by forming a quadrupole field to confine the ions radially while accelerating them axially, leading to denser beams with less dispersion. Soviet development of the RFQ has been carried out by several institutes of the Academy of Sciences and the universities: the Institute of Theoretical and Experimental Physics, the Institute of High-Energy Physics, the Nuclear Research Institute, the Radiotechnical Institute, the Khar'kov Physico-Technical Institute, and the Moscow Engineering Physics Institute.[19] V.V. Vladimirskiy and I. M. Kapchinskiy of the Institute of Theoretical and Experimental Physics hold a Soviet patent for the RFQ.[20]

Soviet development of induction linear ion accelerators (ion linacs) has been remarkable in its steady pursuit of such goals as high current levels, low injection energy, reliability, efficiency, reduced size and simplicity of operation. These criteria provide the practical means of approaching the goal of neutral beam propagation in space.

VI. The electromagnetic energy beams

The basic advantage of electromagnetic energy carriers, such as laser beams, over particle beams, is that their generation and propagation mechanisms are much better understood. Consequently, the development of high-energy lasers is more advanced and they figure more prominently in Western discussions of directed-energy weapons. Since the Soviets do not discuss these weapons, one cannot assess the corresponding Soviet attitudes. However, the level of effort apparent in Soviet particle beam research is not lower than that dedicated to high-energy laser research. Soviet research in high energy lasers has been intensive since the mid-1960s, including practically all the known lasing species and pumping methods. The Soviets have probably placed greater emphasis on electron-beam pumps for lasers than did US laser researchers. An early impetus was provided by Basov's laser fusion programme.

The closest approach to the laser weapon concept in Soviet technical literature can be found in the work of F. V. Bunkin of the Lebedev Physics Institute. Bunkin has been participating in a wide range of high-energy laser research projects, including systematic studies of recombination plasma lasers, heating and ignition of metals in air by intense CO_2 laser beams and non-linear effects of laser beam propagation. One of Bunkin's interests has been the progress towards ever shorter wavelengths at high laser output energy.[21]   In this connection, Bunkin published an evaluation of the nuclear pumped X-ray laser concept, postulating a 0.3-mm copper or brass wire, 1 m long, as an example of the lasing medium, and concluding that the concept was theoretically feasible.[22]   Bunkin also thinks that X-ray and gamma-ray lasers can be pumped by relativistic particle beams utilizing the crystal channelling effect.[23]   This refers to a major Soviet project supported by Velikhov and led by M.A. Kumakhov of the Nuclear Physics Institute of Moscow State University and others.[24]   Kumakhov proposed a channelling radiation X-ray laser tunable within a broad frequency range and characterized by low divergence.   In spite of the obvious limitations of crystal interacting with the electron beam, Kumakhov and others expect very high power output from such a laser.

The free-electron laser (FEL) is a combination of laser and charged particle beam technologies and thus has a potential for retaining the energy advantage of the particle beam.   While the FEL is at present less developed than other types of high-energy lasers, it may surpass them in terms of power output and is therefore a possible basis for a directed-energy weapon.

The Soviets were first to use high-current electron accelerators to pump a FEL: a pulse-line accelerator in 1975, producing 10 MW at 3 cm wavelength with an efficiency of 1.5 per cent.[25]   They were also the first to reach a 50 MW output power level at 3 mm wavelength in a joint effort of the Applied Physics Institute in Gor'kiy and the Institute of High-Current Electronics in Tomsk.[26]   This achievement, however, has been reported amid a marked decrease of publishing activity on Soviet FEL experimentation, suggestive of secrecy imposed on this research area.[27]

The microwave region of electromagnetic beams has been the subject of highly innovative Soviet research.   The main seat of this activity is the Institute of Applied Physics in Gor'kiy, where A.V. Gaponov, the foremost developer of advanced Soviet microwave technology, has been developing gigawatt-to-terawatt devices spanning the millimeter and submillimeter wavelengths, a power and frequency region not accessible to conventional oscillators and amplifiers.   In his drive to produce powerful sources of coherent microwave energy, Gaponov focused on cyclotron resonance masers, an outstanding example of which is the gyrotron. The potential of these devices, perhaps in the advanced version of the plasma-filled gyrotron being developed at the Lebedev Physics Institute under Rukhadze, also places them in the ranks of possible directed-energy weapons.

VII. Conclusions

The Soviet directed-energy R&D programme has several outstanding characteristics. First, the programme started quite early, at the end of the 1950s, which pre-dates serious US interest in this area. Second, a unique emphasis has been placed on pulsed power, regarded as a major component in the national plans for electric energy research and development. This has been without precedent in the West. Third, the research has been pursued on a generic level, in which pulsed power, high-current accelerators, particle beam dynamics and coherent electromagnetic radiation sources are each given the status of a scientific discipline in its own right, subject to systematic theoretical and experimental investigation.[28] In the course of this systematic approach, the Soviets have addressed most of the key scientific issues involved in directed-energy weapon design and created an experimental base capable of supporting a wide range of directed-energy research goals, including the weapons goal. Finally, the programme resulted in a kind of reverse technology transfer, an otherwise rare event in Soviet engineering practice, in which a number of concepts invented or first developed in the USSR were later adopted by US research. While the gyrotron and the radial line accelerator are good examples of this, the most striking are some of the key elements of the neutral beam weapon concept, such as the ion source and the RFQ accelerator. These elements, perhaps better than any other, indicate the weapons capability of the Soviet directed-energy R&D.

The strong position established by the Soviets in the field of directed-energy research and development must also be considered in relation to Soviet defence policies. The R&D effort originated at the time when Soviet leadership supported a comprehensive BMD and it can be argued that the early stress on pulsed power and associated research was a direct result of this policy. If this is true, then Soviet directed-energy technology development was driven by the twin demands of directed-energy weapons and controlled fusion. Such a conclusion would also be consonant with the interests of Velikhov, leader of the Soviet fusion programme and active in pulsed power research relevant to weapons development.

The foregoing discussion strongly suggests that a programme for the development of directed-energy weapons does exist in the USSR. The reversal of Soviet BMD policy that took place in the late 1960s need not have terminated the early weapons development. An important incentive to continue would be the need for the Soviets to demonstrate the technical feasibility of the directed-energy weapon concepts. The exceptional theoretical potential of these weapon demands a resolution of the problem of their practicality even at the cost of considerable and prolonged effort. The high-energy densities involved and the difficulty in scaling down experimental parameters, typical of this field, require the construction of large and expensive experimental facilities. The uncertainties inherent in a host of scientific assumptions underlying the theory of directed energy lead to extensive experimental testing programmes. Finally, the lack of adequate computer simulation support in the Soviet Union imposes further demands for more experimental equipment and empirical testing.

The Academy of Sciences can be expected to cope with these difficulties, probably with support from military research institutions. However, the Academy would retain the scientific leadership of this effort because of its scientific excellence, the accumulated expertise and the pioneering researchers of the field, all of whom are members of the Academy. The Academy's involvement would follow the precedent of the late 1940s, when it was entrusted with the development of nuclear weapons.

While it is not possible to estimate at this time the status of Soviet directed-energy weapons development, the long duration of the R&D effort, the systematic acquisition of theoretical knowledge and the impressive series of ingenious and effective devices lead one to expect that the Academy is capable of creating the necessary technology in the near term to attempt a technical proof of feasibility of these weapons. This would involve a basic breadboard device demonstrating its kill capability in a simplified model environment.

The timing of the technical proof of feasibility, of course, differs for the various types of directed-energy weapons; it may be expected earlier with some versions and missions of the laser weapon, and in those cases it might have already been demonstrated in the USSR.

However, such a proof, if it is possible at all, is a long way from an operational proof of feasibility of a prototype weapon pitted against realistic targets, and is further still from testing complete weapon systems mounted on their intended platforms in a realistic environment. The development of directed-energy weapons beyond the milestone of technical feasibility proof is no longer within the capabilities of the Soviet Academy of Sciences and must include massive industrial participation. But the range of the necessary new, advanced technologies may also exceed the more traditional capabilities of the defence industry. Soviet military R&D, characterized by a conservative style, may not be readily compatible with the required extensive industrial innovation effort and its broad technology ramifications. Most importantly, the poor record of co-operation between the Academy of Sciences and the industry will pose a further, serious obstacle in the course of this development.

Effective translation of the results of the Academy's research into series production has been the perennial Achilles' heel of Soviet industry. Since the Academy of Sciences has been the principal developer of advanced technologies in the Soviet Union, its problems with disseminating these technologies throughout the industry have been one of the major causes of the backward Soviet technology base. Computer technology is a critical example of this situation, with a considerable potential impact on directed-energy weapons. For the past several years, Velikhov has been reorganizing the Academy in an attempt at a radical improvement of Soviet computer technology. In view of Velikhov's past involvement with directed-energy development, his recent efforts may well be due, in part, to the recognition of the crucial role computers play in the development and operation of all aspects of advanced BMD systems.

Thus, in addition to their intrinsic scientific and engineering uncertainties, production of directed-energy weapons in the USSR must overcome the general problems of an inadequate technology base, and the severe qualitative and quantitative underdevelopment of Soviet computer technology, in particular.

The enduring Soviet interest in directed energy may drive their directed-energy weapons programme forward regardless of these problems. If so, the latter may be expected to play a major role in the selection of particular technological solutions used in the approach to these weapons and in the overall design of Soviet advanced BMD systems. The most probable effect of the industrial constraints will be the tendency to minimize the need for large-scale computer-based integration of the system. This will put a premium on fewer, perhaps larger, and more autonomous weapon platforms. As is the case with many other areas of Soviet advanced technology, the practical realization of Soviet achievements in directed-energy R&D may ultimately depend on the success of the various reform initiatives that are being carried out at this time by the Soviet leadership.

Notes and references

1   Simon Kassel is a senior staff member of the National Security Division of the RAND Corporation in Washington, DC. Since joining RAND in 1969, he has been the project leader of research sponsored by the Director's Office, Defense Advanced Research Projects Agency, Department of Defense, in comparative analysis of Soviet applied science and engineering, evaluation of international trends in science, and studies of Soviet science policy and decision-making. For the past decade, he has specialized in directed-energy issues involving pulsed power and high-power laser, particle-beam and microwave development. The author of numerous publications on Soviet directed-energy R&D and the development potential of Soviet advanced technology, he has participated in research projects of the Departments of Defense and Energy, and the National Science Foundation, involving the organization of Soviet R&D and studies of Soviet science manpower training and utilization.

2   A comprehensive account of this kind can be found in Stevens, S., 'The Soviet BMD Program', in: *Ballistic Missile Defense,* eds A.B. Carter and D.N. Schwartz, The Brookings Institution, Washington, DC, 1984, p. 182 et seq. See also *Soviet Strategic Defense Programs*, Department of Defense and Department of State, Oct. 1985.

3   *Pravda*, 11 Feb. 1967, p. 3. A superb analysis of the development of Soviet BMD policy and its changes over time can be found in Parrott, B., *The Soviet Union and Ballistic Missile Defense,* The John Hopkins School of Advanced International Studies, Washington, DC, forthcoming).

4   Talenskiy, N., Dr. Sc. (Mil.), 'Anti-missile systems and disarmament', *International Affairs*, Moscow, Oct. 1964, pp. 16-18. Major General Nikolay Talenskiy of the Soviet General Staff served after retirement as the arms control and disarmament liaison between the Ministry of Defence and the Academy of Sciences, USSR: Garthoff, R.L., 'BMD and East-West relations', in: *Ballistic Missile Defense* (note 2), p. 292.

5   Arkhipov, M., 'Radiation weapons', *Sovetskiy patriot*, 1 Nov. 1964, p. 3.

6   Andreyev, O., 'Possible military and other uses of lasers', *Voyennoye znanye*, Feb. 1965, p. 39; quoted in: Lambeth, B.S., *The Soviet Union and the Strategic Defense Initiative: Preliminary Findings and Impressions,* The Rand Corporation, N-2482-AF, June 1986, p. 20.

7   Kassel, S., *Pulsed-Power Research and Development in the USSR,* The Rand Corporation, R-2212-ARPA, May 1978, p. 3.

8 Kassel (note 7), pp. 115, 116.

9 Kassel (note 7), pp. vii, 8, 16.

10 Velikhov, Y.P., et al., 'The use of pulsed MHD Generators in geophysical research and earthquake prediction', Kurchatov Institute of Atomic Energy, Moscow, 1975, 16 pp.

11 Hutchinson, R., *Janes's Defence Weekly*, vol. 4, no. 16 (19 Oct. 1985), pp. 837.

12 Kassel (note 7), p. 115.

13 Pavlovskiy, A.I., et al., 'LIU-10 High-Power Electron Accelerator', *Doklady Akademii Nauk SSSR*, vol. 250, no. 5, 1980, p. 1118.

14 *Byulleten'izobreteniy*, no. 34, 1971, p. 223.

15 Kassel, S. and Hendricks, C.D., *High-Current Particle Beams. I. The Western USSR Research Groups*, The Rand Corporation, R-1552-ARPA, Apr. 1975, p. 7.

16 Kassel and Hendricks (note 15), p. 2.

17 Kassel and Hendricks (note 15), p. 3.

18 Wells, N., *The Development of High-Intensity Negative Ion Sources and Beams in the USSR*, The Rand Corporation, R-2816-ARPA, Nov. 1981.

19 Wells, N., *Radio Frequency Quadrupole and Alternating Phase Focusing Methods Used in Proton Linear Accelerator Technology in the USSR*, The Rand Corporation, R-3141-ARPA, Jan. 1985..

20 *Physics Today*, Aug. 1983, p. 19.

21 Bunkin, V.F., Derzhiyev, V.I. and Yakovlenko, S.I., 'Prospects for light amplification in the far ultraviolet', *Kvantovaya elektronika*, vol. 8, no. 8, 1981, p. 1621.

22 Bunkin, V.F., Derzhiyev, D.I. and Yakovlenko, S.I., 'Requirements for pumping an X-ray laser by an ionizing source', *Kvantovaya elektronika*, vol. 8, no. 7, 1981, pp. 1606; Bunkin noted Clarence Robinson's article on Livermore's X-ray laser (*Aviation Week & Space Technology*, no. 8, 1981, p. 25), complaining that the information was far too insufficient to judge the reliability of the experiment.

23 Bunkin (note 21), p. 1647.

24 Kassel, S., *Soviet Research on Crystal Channeling of Charged Particle Beams*, The Rand Corporation, R-3224-ARPA, Mar. 1985.

25 Tkach, Y.V., Faynberg, Y.B., et al., Physico-Technical Institute, Khar'kov, *Zhurnal eksperimental'noy i teoreticheskoy fiziki*, Pis'ma, vol. 22, no. 3, 1975, p. 136.

[26]   Denisov, G.G., et al., *Int. Journal of Infrared and Millimeter Waves*, vol. 5, no. 10, 1984, p. 1389.

[27]   Kassel, S., *Soviet Free-Electron Laser Research*, The Rand Corporation, R-3259-ARPA, May 1985, p. 40.

[28]   A typical example is the comprehensive study of explosive electron emission at the cathode surface, carried on for many years by Mesyats at the Institute of Atmospheric Optics and later at the Institute of High-Current Electronics in Tomsk. Mesyats has made a science of exploring this basic mechanism of electron beam formation.

Paper 5. Technologies of ballistic missile defence

Thomas H. Johnson[1]
Scientific Research Laboratories, USMA, West Point, NY 10996, USA

I. Introduction

The first three sections of this paper will discuss certain US weapon systems and their capabilities for performing ballistic missile defence (BMD) functions. The remainder of the paper discusses relevant portions of the programme of the US Strategic Defense Initiative (SDI).

II. Existing systems

The United States has no operational system dedicated to ballistic missile defence. In 1976 the United States briefly deployed a system of interceptor missiles and radars as prescribed by the ABM Treaty of 1972. This system, in accordance with the Treaty, was located at a single site, Grand Forks, North Dakota. A few months after it was declared operational, the site was shut down. It has since been thoroughly deactivated. The United States currently maintains two designated BMD test ranges: Kwajalein Missile Range, on Kwajalein Atoll in the Pacific; and White Sands Missile Range, at White Sands, New Mexico. At both sites ABM interceptor missiles, ABM radars and ABM launchers are tested in compliance with Treaty restrictions.

III. New technologies

Anti-satellite (ASAT) weapons

The United States currently has a single ASAT system which is still undergoing advanced testing. This system has been under engineering development since the 1970s, and has been tested once (successfully) against a satellite target.

The US ASAT destroys satellites by direct impact. It is launched from F-15 aircraft at moderately high altitude, ascending during its boost phase by inertial guidance directly toward the target satellite's orbit. It closes with the target by long-wave infra-red homing. As currently planned, the operational version of this weapon will be based only at a single site.

This particular ASAT system would have virtually no utility in a BMD role. It is not designed to operate in the atmosphere and so would be incapable of boost or terminal phase intercepts. Although conceptually it might be possible to perform single intercepts of re-entry vehicles (RVs) near apogee with such homing vehicles, details of cost and performance of the ASAT would make such use unattractive, and operational details (F-15s would have to be deployed at appropriate geographic positions and altitude at attack time) would make it virtually impossible.

Air defence and anti-tactical missiles

The most capable system used by the US for air and tactical missile defence is the Patriot missile system. Patriot was designed as an air defence system, with some residual capability against short-range ballistic missiles. As far as ICBMs are concerned, the Patriot radar does not have sufficient power-aperture to perform the requisite acquisition and tracking. The Patriot missile itself can only engage targets within the atmosphere and is too slow to perform intercepts of ICBMs (as SPRINT was designed to do), even when supplied with adequate radar information. The missile does not have sufficient slant range (combined range and altitude) to perform exo-atmospheric intercepts (as Spartan was designed to do). Thus, the US has no deployed air defence systems with BMD capabilities. Systems based upon the present research effort will be discussed below.

Directed-energy weapons

Lasers. Space-based lasers have received more public attention than any other directed-energy BMD technology. Chemical lasers, particularly the hydrogen fluoride laser, are probably the lasers of choice for this mission, because they have the highest demonstrated power and because they obviate the need for large, heavy electrical power supplies in space. Ground-based lasers must engage boosters over much greater distances; at such ranges, the wavelength of chemical lasers would dictate impractically large optical components. Thus, chemical lasers are not appropriate for ground-based boost-phase kill. After a long, intense debate, experts have agreed that for nominal laser powers, roughly 100 chemical laser battle stations in orbit might be needed to deal with a threat of 1000 ICBMs.

In spite of their superior state of development (compared to other laser candidates), chemical lasers have recently been de-emphasized by the US SDI programme. The most important reason for this change has been the projected development time-scale for the chemical lasers. The NACL chemical laser provided one of the first laser lethality demonstrations when it destroyed a TOW missile in 1978. A larger version of NACL, called MIRACL, was then built and subsequently moved to White Sands missile range, where it recently performed a lethality experiment on an ICBM booster. Although chemical performance has improved somewhat since 1978, this improvement is unlikely to lead to the higher brightnesses needed for space-based chemical systems. A new laser (called Alpha), which would take the first steps towards such improvements, is still under development, and further generations will be needed. Thus, chemical laser technology seems to be in a rather gradual stage of maturation.

A further criticism which has been voiced concerning the chemical lasers is their vulnerability to a direct, defence-suppression attack. A critical element in surviving such an attack is the laser's retarget time. Proposals to protect the lasers with kinetic-energy weapons (KEWs) have been made; whether this can work remains problematic. Moving the laser to higher orbits could increase its survivability, but would require a laser of significantly higher brightness and, by inference, significantly longer development time.

Lasers of wavelengths much shorter than those of the chemical lasers could be based on the ground, their beams relayed towards rising ICBMs by mirrors in space. Some of these mirrors would be predeployed in orbit; others might be launched directly to the required altitude at the initiation of the attack (or 'popped-up', as the jargon has it). Currently laser candidates for this mission require large electric power supplies and so appear to be unrealistic candidates for space basing. The short-wavelength ground-based lasers fall into two disparate categories: free-electron lasers (FELs) and excimer lasers.

Free-electron lasers are currently the highest-priority directed-energy weapons in the US programme. There are two different basic designs for FELs, based upon accelerator technologies known as induction accelerators and radio-frequency (RF) accelerators. Gas lasers, like chemical lasers and excimers, work by depositing large amounts of energy in gases and then extracting some fraction of that energy in a collimated beam. FELs work by accelerating electrons to very high velocities, then causing them to radiate some of their kinetic energy at the proper wavelength and in the proper direction in resonance with light waves in the electron beam.

FELs came to prominence as potential defence weapons in 1985 when it was announced that an FEL experiment at Livermore had produced high radiation power at very high efficiency (that is, efficiency of converting electron energy to radiation energy). If FELs of different design parameters can produce beams of the appropriate wavelength and beam divergence at equivalent efficiencies (or even close), they might be strong competitors for the boost-phase mission, although extremely high powers would be required. More fundamental issues must first be addressed: the wavelength of radiation must be reduced a factor of several thousand from that of the Livermore demonstration, and the resultant beam must have a high quality. In short, several questions of basic physics have to be answered before experiments at full scale became appropriate. FEL theory currently supports the optimistic extrapolations, but the high-risk quality of the enterprise stems from the shortage of empirical support for scaling. The RF FEL, unlike the induction FEL and the excimers, could conceivably be employed space-based (as well as ground-based) because of its relative compactness.

Excimer lasers, invented in the mid-1970s, were the first lasers to operate at visible and ultraviolet wavelengths with efficiencies over 1 per cent. Their efficiencies remain low compared to both chemical lasers and FELs. The excimers of interest for strategic defence involve the combination of noble gases (xenon, krypton) with halogens (chlorine, fluorine). Unlike both the chemical lasers and the FELs, some excimers are not intended to damage targets by depositing their energy as heat over relatively long pulses (significant fractions of a second); rather, their higher-energy photons, deposited in very short pulses, are expected to cause damage by impact, driving a pressure wave into the target much as a projectile might do. Other designs use repetitive, very short pulses to accomplish thermal kills. Energy, rather than power, is the measure of the excimers' effectiveness.

Large, high-energy excimer lasers have been developed for the Department of Energy's inertial fusion research programme and by the

Defense Advanced Research Projects Agency (DARPA). Such a laser, currently operating at Los Alamos, is almost the size of the largest single module of an excimer that is sensible to build. A strategic defence excimer system would combine the beams from a large number of such modules to produce the requisite beam energy. Although some physics issues in the beam combination remain to be demonstrated at full scale, the main issues in excimer lasers are the formidable engineering and optics problems associated with combining the many apertures while retaining beam quality without losing massively in efficiency.

Three additional problems face both types of ground-based laser (GBL). Because they must propagate up through the dense atmosphere to reach their relay mirrors, the beams of GBLs must pass through many random cells of atmospheric turbulence, which has the effect of a distorting lens in spoiling the optical quality of the beams and hence reducing their destructive power on targets. If the exact nature of the distortion is known, however, it is possible in theory to distort the surface of the laser's final mirror to compensate exactly for the effect of the atmospheric turbulence, a technique known as adaptive optics. Experiments have demonstrated that this technique does work. However, those experiments were conducted in a relatively clear atmosphere and with low-power lasers. More adverse atmospheric conditions and high laser power remain to be studied in detail over the next few years.

The second GBL problem is related to the first in that it is caused by weather conditions: none of these lasers is capable of penetrating cloud layers without depositing all or most of its energy there. Hence, a GBL strategic defence would have to have enough sites to guarantee that enough lasers would be cloud-free to perform the mission. Studies of weather statistics show that several sites could provide better than a 98 per cent probability of having one site cloud-free. Thus, if one laser facility were sufficient in destructive power, 10 sites or fewer would probably suffice for the mission.

The third problem is survivability. The GBL's space relay mirrors, although smaller and cheaper than space-based laser (SBL) battle stations, are susceptible to attack. Perhaps more important, the ground system comprises a limited number of large laser facilities, all with known locations. Thus they are susceptible to direct attack by submarine-launched ballistic missiles (SLBMs), submarine-launched cruise missiles (SLCMs) and sabotage. So any ICBM defence which depends strongly on GBLs must also be effective against SLBMs and SLCMs;

and very effective means of preventing determined attempts at sabotage by enemy agents must be developed.

Lasers have been considered primarily as boost-phase weapons because the boosters are considerably softer targets than the re-entry vehicles they carry. The real lethality of various laser candidates is still a subject of intensive research. The question is not whether laser beams can actually destroy targets of interest, but rather what wavelengths, fluxes and pulse-lengths cause how much damage against what kinds of targets. The available hard data on this subject is still rather spotty; the products of the research will be important in dictating the requirements of lasers for particular missions. It is thought unlikely that lasers will be effective mid-course weapons because re-entry vehicles are very hard targets and the engagement ranges are still very long: the combination prescribes extremely high brightnesses. Retarget time is again an issue.

One final laser weapon deserves mention, but cannot be described in detail: the X-ray laser. The Department of Energy is conducting a research programme to use a nuclear explosion to produce lasing radiation in X-rays; it has been announced that the scientific goal of producing such radiation has been achieved. Such a laser would self-destruct as it fired, and would presumably launch multiple beams towards their targets. The principal attractiveness of the X-ray laser lies in the extreme lethality of high brightness beams of X-rays at great distances. One way to look at the laser is to consider it as a means of taking the copious X-radiation normally produced isotropically by the bomb and focusing all that radiation into narrow beams. An X-ray laser's operational regime would be limited to space and the upper atmosphere, since X-rays do not propagate far in the atmosphere. Systems studies typically presume that the enhancement of X-ray fluence within the beams (compared to what would arrive from a 'normal' bomb) should be about a factor of one million or more. The status of the research has been roughly described by Dr George Miller of Livermore: '...what we have not proven is whether you can make a militarily useful X-ray laser. It's a research programme where a lot of physics and engineering issues are still being examined...There's a lot of work to do, not the least of which is actually designing the system, not just the laser itself.[2] A successfully developed X-ray laser system might be used to perform mid-course discrimination (if large numbers of beams per laser were possible), mid-course kill (if very high brightness were possible) or terminal kill in the very high atmosphere (as an adjunct to other terminal-phase systems). Virtually all analysts agree, however, that a working X-ray laser, lifted to the upper atmosphere by a rocket, could be an effective anti-satellite weapon. Whether such a laser could actually assist in boost-phase kill depends on complex systems issues (the available engagement times being very short), and its usefulness is still regarded as problematic.

Particle beams. Beams of charged particles (such as electrons) will not propagate in space, because the individual particles, having like charges, strongly repel one another. This repulsion can be stabilized, at least temporarily, in propagation within the atmosphere; but neither charged nor neutral beams propagate very far within the dense atmosphere before being scattered and broken up.

Thus, the largest component of particle beam work is a research programme in the generation and propagation of neutral beams to be deployed in space. These beams are typically thought of as comprising atoms of deuterium (heavy hydrogen) at very high particle energies and rather low currents. The atoms are accelerated through electric fields as negative ions (with an extra electron attached); then the electron is stripped off in passage through a gas cell, leaving a beam of neutral atoms. The advantage of the beams as weapons is that their target penetration is so high that it is virtually impossible to shield against them. The problem with them is that, at the low currents known to be plausible, the beams have to have very long dwell times on target to deposit lethal fluences of energy.

Neutral beam accelerators would reside in large space-borne battle stations, from which they could address boosters or the 'buses' containing RVs above about 120 km and re-entry vehicles throughout mid-course. The important issues are: the divergence angle of the beam tracking and kill assessment; and the retarget time, since the beam must be slewed by magnetic refocusing lenses. Because of limitations of beam brightness, current neutral beam concepts are being designed for mid-course discrimination missions: the deposited particles will cause the RVs to radiate in amounts proportional to their masses, thus making the heavy RVs stand out from decoys. This mission is less ambitious in terms of brightness and kill assessment problems, but it would require an extensive constellation of satellite-borne detectors to collect and process the discrimination data. Attainable divergence angles will apparently require neutral beam stations to remain at rather low orbits, which raises the problem of survivability against direct attack, especially if retarget time is long.

Accelerator development and laboratory experiments at fairly small scale have been conducted for some years, with impressive gains in accelerator technology (using important Soviet innovations). The programme is about to expand to large-scale lab experiments to scale up the physics results to meet test requirements on the beams. Actual space testing could then follow.

Advanced kinetic-energy weapons

Kinetic-energy weapons fall into four categories, with somewhat different performance requirements, based on the regime of the ICBM trajectory where intercept is planned to occur. Intercepts by KEWs in the boost phase and pre-apogee in the mid-course must be handled by projectiles based on satellites, since ground-based interceptors would not have time (nor range) to reach their targets; these interceptors are generally known as space-based kinetic kill vehicles (SBKKV). The architecture of their employment, either as a primary boost-phase kill system or to protect other space-based assets, rather rigidly prescribes the requisite performance characteristics. For example, boost-phase intercepts require very high velocity vehicles with advanced sensors to reach their targets in time; they must be very light, and housed in satellites in relatively small numbers to reduce their value to direct attack. Mid-course intercepts by SBKKVs require large numbers of satellites to provide reasonable ranges; most important, they require solution of the mid-course discrimination problem.

Intercept in the late mid-course might be performed by ground-based miniature homing rockets lofted on larger boosters or individual rockets. Generally, optical homing sensors are envisioned for this mission, since the homing will be done outside the atmosphere and the sensor will generally not have to look into the atmospheric background radiation. The single-rocket approach is taken by the SDI programme Exoatmospheric Reentry Vehicle Interception System (ERIS), which aims to put very small interceptors on rockets of size comparable to field artillery missiles.

Intercept high in the atmosphere will similarly require a high-velocity missile with fast response and the capability to move relatively long distances off its trajectory as it nears its target (high 'lateral direct range'). Intercepts at the higher altitudes may use optical/infra-red sensors, but at low altitudes radar homing may be more appropriate, because of background radiation, nuclear effects and the difficulty of keeping optical sensors cool during high-velocity flight in the atmosphere. This element of the SDI programme is called HEDI (High Endoatmospheric Defense Interceptor).

Finally, terminal intercepts might be performed by smaller, highly agile missiles guided by radar homing. Such an experimental vehicle destroyed a stationary target, then a moving one, in 1986; the project is designated Flexible, Lightweight, Agile Guided Experiment (FLAGE).

All these programmes require significant experimental progress in reducing the weight and improving the performance of the interceptor vehicles and their boosters. Key areas of improvement include fast-reaction manoeuvring engines; new structural techniques and materials; miniaturized avionics; and higher rocket mass fractions, specific impulse and chamber pressures. All the vehicles must be capable of homing within, at most, a few metres of their targets; this by itself is a formidable requirement in terms of technologies for sensors, guidance and control, and direct propulsion.

Assuming that all these engineering developments meet their objectives, two dominant problems remain to make the KEW effective. Outside the atmosphere, the mid-course discrimination problem must be solved; no matter how inexpensive the SBKKV and ERIS can be made, they will not be able to address all of the hundreds of thousands of objects which they may face. Inside the atmosphere, KEWs will have to find a way to deal with manoeuvring re-entry vehicles (MARVs). These vehicles use the high velocity of the terminal RV to accomplish extremely high-G manoeuvres when they reach the dense atmosphere, simply by extending fins into the surrounding flow. MARVs which execute such manoeuvres to evade intercept may or may not have sufficient accuracy to attack hard military targets, but they certainly present a strong threat to cities or other extended or soft targets. Their velocities, their tight turns, and the unpredictability of the direction of such turns make such intercept by KEWs entirely problematic.

Surveillance, acquisition, tracking and kill assessment

These functions are usually grouped together because they involve related technologies, although the technologies may be employed in vastly different ways.

Surveillance of ICBM launches and tracking of boosters for intercept require major improvements in the existing systems that provide early warning of ICBM launches. The most important of these systems are the high altitude satellites which detect the strong infra-red signals emitted by the boosters' fiery plumes. Discovering that strong infra-red signals exist is not difficult; the problem is in localizing and tracking them to very high accuracy, over long distances, and at angles which may bring in noise signals from the earth's own radiation.

Once the booster has ceased to burn and the re-entry vehicles begin to emerge from the bus, tracking becomes more difficult. Bus, RVs and decoys do emit infra-red signals, but very weak ones. They must be isolated and tracked to very high accuracy at ranges of many hundreds of kilometres, a task beyond anything yet publicly demonstrated. Then, as mentioned above, the problem becomes discriminating the RVs from their decoys.

Considerable effort has gone into examining the infra-red and visible (reflected sunlight) signals from RVs to determine whether discrimination can be accomplished by passive observation of this radiation; the present consensus is that it cannot. This leaves two other techniques: active observation, in which radar waves or optical radiation (from lasers) are bounced off the target, in effect taking its photograph; and interactive means, in which energy or momentum is transferred to the RV (by lasers, particle beams or even pellets) and the response to that deposition observed.

Whether active means can be made to work remains problematic, although considerable effort at imaging radars has gone on in the past and continues. Work now seems to be concentrating on the interactive means, since there is hardly any question of whether they can actually perform the discrimination: if, for instance, a laser can blow off substantial vapour from the target's surface, the target's recoil will reveal its mass and thereby its identity. The question regarding interactive means, then, is whether they can be made practical.

The SDI programme is pursuing laser radar technologies; this work is largely a by-product of research on pointing and tracking for directed-energy weapons. Particulars of the status and requirements of laser transmitters remain classified. The work thus far has been research and technology development; no integration experiments are currently planned.

For terminal phase intercepts, some imaging of targets will also be useful, to improve battle management and to deal with future re-entry vehicles whose performance has been improved to neutralize the effects of the defence (so-called 'responsive threat' vehicles). Research on advanced high-frequency radars to perform this function is also under way.

The US, like the USSR, maintains a system of wide-cone phased-array radars for early warning and attack assessment, including warning of SLBM attacks. All the US radars are located and oriented in accordance with the provisions of the ABM Treaty. Thus, they are geographically incapable of performing battle management functions for BMD. Although the precise frequencies of these radars are not published, it should be noted that the physical requirements for efficient long-range detection and for battle management are different. To optimize acquisition, one would design a radar in the ultra-high frequency (UHF) band, where the wavelength of the radiation is on the same scale as the size of the object (re-entry vehicle) to be detected. On the other hand, battle management radars should be designed to operate at frequencies as high as are practically possible (X-band or higher), to be able to penetrate the ionization caused by the atmospheric nuclear detonations.

For lasers and KEW, kill assessment means looking at the target again to see whether it has broken apart or suffered severe structural damage. Practical considerations (numbers of picture-takers and their locations, choice of picture-taking technology and data management) dominate the analysis. For neutral beams the problem could be more difficult: depositing their energy deep within their targets, the beams could leave no visible evidence of success, even when the attacking warheads have been destroyed. The technical problems of kill assessment, like those of discrimination, are still under laboratory study.

IV. Conclusion

Several technological areas addressed here will require significant advances before the goals of SDI can be realized—particularly power sources in space, inexpensive means of putting large payloads in orbit, and command and control systems. The first two are being broadly pursued internationally on their own merits; the third should not raise ABM Treaty issues independent of the weapons being controlled.

The US has no deployed systems of any kind with technical capabilities that pose Treaty problems. None of the technologies currently within the research programme involves inherent contradictions with the language of the Treaty, so long as tests continue, as at present, within the Treaty's provisions.

Notes and references

1   Thomas H. Johnson is Professor of Applied Physics and Director, Science Research Laboratory, US Military Academy, West Point, New York. He has a Ph.D. in applied physics. Previous appointments include: US Air Force Weapons Lab; US Defense Nuclear Agency; Lawrence Livermore National Laboratory; and US Military Academy. 1981-82 he was Special Assistant to the President's Science Advisor and in 1986 received the IEEE Donald Fink Prize. Currently he is Lieutenant Colonel in the US Army

2   George Miller, Associate Director, Lawrence Livermore National Laboratory, quoted in 'Experts cast doubt on X-ray lasers', *Science*, vol. 230, no. 647, 1985.

Part IV. ABM Treaty issues—US and Soviet views on the Treaty and compliance questions

Paper 6. Issues related to current US and Soviet views of the Treaty: a Soviet jurist's perspective

Vladlen S. Vereshchetin[1]

Institute of State and Law, USSR Academy of Sciences, Moscow

I. Reykjavik and after

The perspective of deep reductions of nuclear weapons and their eventual elimination that emerged at Reykjavik called, naturally enough, for a simultaneous agreement to ensure that a lethal channel of the arms race would not be left open—that of space arms. Each party should have guarantees that its security will in no way be jeopardized; that stability and parity will remain intact; and that no situation will arise wrought with elements of uncertainty and unpredictability.

Believing that under the circumstances the Anti-Ballistic Missile (ABM) Treaty does not lose but rather gains in importance, the Soviet side proposed at Reykjavik a number of measures aimed at strengthening the Treaty. The Soviet proposal envisages neither the revision nor the substitution of the Treaty. It is aimed at confirming the strength and significance of its provisions and at consolidating the regime established by it.

The substance of the Soviet proposal on the subject handed over to the President of the United States was formulated in the following way in the text:

The USSR and the USA would commit themselves not to use for ten years their right to withdraw from the ABM Treaty of unlimited duration and to adhere closely to all of its provisions during that period. All tests of space elements of ballistic missile defence in outer space are banned with the exception of research and tests conducted in laboratories.[2]

The proposal for a non-withdrawal period of 10 years did not imply the substitution of the treaty of unlimited duration by a new 10-year-long accord. This would serve only as an additional guarantee to each of the parties that for the coming decade, at least, while nuclear weapons were being eliminated, no new weaponry would take their place. Alongside the commitment to adhere strictly during that decade to all the provisions of the Treaty, the Soviet proposal contained yet another important element. In essence the Soviet Union agreed to an interpretation of Article V of the Treaty prohibiting the development, testing and deployment of space-based ABM systems or components. The interpretation was that the ban on research and development would not be extended to work conducted within the framework of laboratories. Thereby the Soviet Union offered a compromise: in view of the attitude of the US President towards the Strategic Defense Initiative (SDI) and his repeated statements that SDI was no more than a research programme, the Soviet Union was ready to recognize the legal admissibility of US research and testing activities under that programme within a laboratory framework. This was a retreat from the former Soviet posture during the Geneva talks when the USSR insisted on completely prohibiting the creation (including research activities) of space-based ABM means, a retreat made for the sake of achieving a reasonable compromise on the issue.[3]

At the same time this would have been a step forward towards strengthening the ABM Treaty—the proposal not only took into consideration the existing realities, but would also bring about a uniform interpretation by both parties of one of the important provisions of the Treaty.

The flexibility of the Soviet stand on the issue manifested itself also during the subsequent Vienna meeting of the Soviet Minister of Foreign Affairs with the US Secretary of State. The Soviet Union suggested that top-level negotiations be started in the nearest future to define which activities were allowed by the ABM Treaty and which were not. As the development of anti-satellite weapons might result in evasion of the ban on the creation of space-based ABM weapons, the Soviet Union proposed that mutually acceptable decisions on the prohibition of anti-satellite weaponry should also be sought.[4]

The adoption of the Soviet proposal on the strict observance of all the provisions of the ABM Treaty for the coming decade at least and on a uniform interpretation of one of its key provisions would not impose a ban on research and testing under the SDI programme conducted in laboratory conditions. As to a more remote perspective, the Soviet Union proposed special negotiations with a view to arriving at mutually acceptable decisions on the future course of action with regard to these matters.

What was the US reaction to these proposals? Soviet Minister of Foreign Affairs E.A. Shevardnadze summed up the US stance in this way: '...we were invited to sanction the deployment of space weapons and to sign the death sentence of that treaty, only staying off the execution for 10 years'.[5]

At the Reykjavik meeting the US President suggested the substitution of the ABM Treaty by some new agreement controlling the creation and deployment of space-based ABM systems. At the press conference arranged immediately after the Reykjavik meeting M.S. Gorbachev pointed out that 'the President kept insisting to the very end that the US had the right to research and test everything related to the Strategic Defense Initiative, not only in laboratories, but outside of their confines, outer space included'.[6]

It is evident that such demands are incompatible with the observance of the ABM Treaty, which quite definitely prohibits the development, testing and deployment of space-based ABM systems and components. It was a claim aimed at securing an official Soviet agreement to the notorious 'broad' interpretation of the Treaty, the meaning and intentions of which will be examined in this paper. However, even such a basically revised version of the Treaty would not meet the wishes of the US Administration in the future, and Washington announced beforehand that it contemplated withdrawal from it in 10 years' time to legitimize the beginning of deployment of an ABM system with space-based elements. Moreover, the Soviet side was invited to undersign such a perspective. That was an invitation to sign a 'death sentence' to a treaty that is universally looked upon as the corner-stone of the entire structure of international agreements on arms limitation and reduction.

Immediately after the Reykjavik meeting the President, the Secretary of State, the National Security Adviser and other officials of the US Administration repeatedly alleged that the USA was insisting on 'strict observance' of the ABM Treaty. They also alleged that the Soviet proposal at Reykjavik was aimed at revising the Treaty so as to make it much more restrictive than was agreed upon by the parties at the time it was signed and in comparison with the so-called 'narrow' interpretation.

At the same time, the US statesmen mentioned above asserted that in their view 'strict observance of all the provisions of the ABM Treaty' was compatible with not only research and development of space-based ABM systems, but also with related tests. That stance was also upheld by the US President at Reykjavik.

In view of the above, two questions are pertinent here: what meaning does the US Administration put into the phrase 'strict observance of all the provisions of the Treaty' and what is the intention behind and the meaning of its narrow interpretation? As is known there is no article-by-article interpretation of the whole Treaty agreed upon by the parties to it. There are 'agreed statements' and 'common understandings' covering specific articles of the Treaty adopted by the parties when it was signed. Some issues related to the terminology of the Treaty were subsequently discussed by the Standing Consultative Commission. No joint 'agreed statements' or 'common understandings' were agreed upon with regard to Article V, para. I, of the Treaty—crucial for the present interpretation of the instrument.

A review of official pronouncements of the US Administration reveals the presence of at least five interpretations or 'practical policies'—all at variance with one another. It is noteworthy that four of these came into being after the SDI programme had been announced.

II. Pre-SDI interpretation

In 1972, when the Treaty was brought before the Senate pending its ratification, high ranking officials of the Administration more than once stated that the provisions of Article V banning the development, testing and deployment of sea-based, air-based, space-based or mobile land-based ABM systems or components take effect 'at that part of the development process where field testing is initiated on either a prototype or breadboard model'.[7] In other words, the US Administration, according to that clarification, considered that the division line between banned and allowed research on sea-based, air-based, space-based and mobile land-based ABM systems and components runs between laboratory work and field testing. The following reason was given to lend credence and substance to such a divide: the laboratory stage does not allow for reliable verification by national technical means. Consequently, verification would present severe problems at the earlier phases of the creation of ABM systems and components.

It should be noted here that in 1985 and 1986 the Soviet Union, advancing at the Geneva talks its proposal on the complete ban on the creation of space strike weapons (including laboratory work for the purpose), suggested that the verification issue should be settled by opening up the respective laboratories for on-site inspections.[8]

In the light of the present interpretation controversy it is essential to note yet another point: according to documents published in 1972 when the Treaty was in its ratification stage, both the US Administration and the Congress quite clearly considered that the Treaty banned extra-laboratory 'field' testing of space-based ABM systems and components. Did that refer only to the ABM systems and components that already existed at the time or to future means based on new physical and technological principles as well? The statements made in the US Senate by Melvin Laird, then Secretary of Defense, General Bruce Palmer, Acting Chief of Staff of the US Army, General John D. Ryan, Chief of Staff of the US Air Force and by other representatives of the US Administration, as well as the questions raised by senators and answers given to them during the hearings, indicate beyond any doubt that both the US Administration and the Senate realized quite clearly that the ban referred both to 'current' and to future technologies that did not exist at the time the ABM Treaty was signed. Moreover, that was the reason why Senator James Buckley, one of the two senators who voted against the ratification of the Treaty, cast a 'no' vote on the subject.[9] No other interpretation of the provisions of the Treaty was offered when it came up before the Congress during the ratification hearings.

John B. Rhinelander, legal adviser to the US Delegation at the SALT I talks, answering questions by Mr Dante B. Fascell, Chairman of the House Committee on Foreign Affairs, wrote:

To the best of my knowledge all Administrations since May 26, 1972, when the treaty was signed (Nixon, Ford, Carter and Reagan) until October 6, 1985, when the 'reinterpretation' was first disclosed on television, had interpreted the ABM Treaty as covering both 'current' (or traditional) ABM systems and components as defined in Article II (I) and 'future' or 'exotic' technology, such as lasers. This was clearly stated in 1972 by the Nixon Administration and understood by the Senate.[10]

Characteristically, way back in 1984 when the current Arms Control Impact Statement prepared by the Arms Control and Disarmament Agency director was submitted to the Congress, Reagan repeated, as he had done on several occasions previously, exactly this interpretation of the ABM Treaty. The statement says:

The ABM Treaty prohibition on development, testing and deployment of space-based ABM systems, or components for such systems, applies to directed energy technology (or any other technology) used for this purpose. Thus, when such directed energy programs enter the field of testing phase, they become constrained by these ABM Treaty obligations.[11]

III. Reinterpretation begins

The declaration of the SDI programme by its very nature clearly contravenes the ABM Treaty because it represents an attempt to set up an 'anti-missile shield' covering the entire territory of the United States, which is directly and explicitly prohibited by Article I of the Treaty.[12] SDI thus presented the Administration with the problem of how to find in the text of the Treaty omissions, loopholes, language of insufficient clarity and terms that had not been defined in the Treaty—the so-called grey zones—in an attempt to substantiate its claim that the SDI programme is compatible with the ABM Treaty. It is absolutely clear that this could not be done by using procedures provided by the terms of the Treaty, that is, through the Standing Consultative Commission. Conversely, the task was made easier by piling up 'narrow' and 'broad' interpretations, 'practical policies' and the like so that the uninitiated would be befuddled and lost in the maze of such reasoning.

Even before the 'new interpretation' was launched there appeared a new version, a modification of sorts, of the former official interpretation of the Treaty.[13] Moreover, at first the new version existed side by side with the old interpretation in a number of executive documents (in the reports, among others, to the US Congress by the Arms Control and Disarmament Agency). At that time nothing was said overtly about such an interpretation change, but in point of fact the revision was already well under way owing to the process of 'squeezing' activities under the SDI programme into this or that provision of the Treaty.

Specifically this stance of the US Administration manifested itself in the report on the Strategic Defense Initiative submitted to the US Congress by the Department of Defense in April 1985.[14]

On the one hand, the report alleges that SDI is being implemented 'in a manner fully consistent with all US treaty obligations'.[15] On the other, when listing the activities under that programme the authors of the report imply that Article V of the Treaty does not obstruct SDI activities because, firstly, that programme is limited only to some preliminary stage in the development and testing of space-based ABM elements, and, secondly, those elements, allegedly, do not fall under the 'components of a system' concept mentioned in Article II of the Treaty.

SDI supporters assert that since the current programme is only a research project and since the decision on the expedience of deploying the system will only be made in the future, the present stage the SDI programme does not violate the ABM Treaty. But in reality what they try to pass off as innocent 'research' is a plan, unprecedented in its scope, cost and concentration of scientific and technological effort, the objective of which is a stage-by-stage creation of what is termed in the Treaty a 'base for the ABM defense of the territory of a country'. Numerous 'technologies', 'devices', 'subcomponents', 'models' and 'mock-ups' are all to become constituent parts or 'building blocks' of that base.

The textual interpretation of the terms used in Article V—'not to develop, test, or deploy' a system or component—renders legitimacy to the conclusion that the Article prohibits not some 'advanced' stages of the development of systems and components, but, in general, their development *ab initio*.

One of the arguments which are used for lending 'legitimacy' to the wide scope of the R&D projects envisaged by SDI is the linguistic discrepancy in the Russian and English texts of Article V of the Treaty. The Russian text uses the word 'sozdanije' (creation); the English text renders the corresponding concept through the word 'development'.[16] However, it would be extremely illogical to use this divergence to legitimize a wide range of research and development projects that are at variance with the aim and object of the Treaty. It would be rather more logical to acknowledge the wisdom of eliminating that linguistic divergence through a proper accord between the parties.

Yet another example is pertinent to our analysis of the US interpretation of the Treaty. The development and testing of the elements of future ABM systems under the SDI programme are claimed to be permitted by the Treaty under the pretext that they are not all termed components (the term used in the Treaty), but 'subcomponents' or 'devices', while testing (directly prohibited by the Treaty) is given the misnomer of 'demonstrations'. This is precisely the terminology used in the report on the SDI programme submitted by the Department of Defense to the US Congress.[17]

It is quite obvious that the word game and the substitution of some terms by others in no way affect the substance of the matter; neither can they change illegal activities into legal ones. This has not escaped, in particular, the attention of US authors who have looked into the problem under review here. 'Demonstration' is the term the Pentagon often uses in describing activities that in the ABM Treaty language are determined as 'tests'. At the same time the narrow understanding of the term components 'ignores the history of the Treaty negotiations and...rests on an extremely limited concept of what constitutes an ABM system...'.[18]

IV. 'New' interpretation

The realization of the SDI programme might call at a certain stage for full-scale tests of space weapons to be followed by integration tests of the space system. So the US Administration initiated in 1985 a 'new interpretation' of the Treaty, the third in a series of such previous revisions. This caused an uproar the world over—the move literally put on its head the settled understanding of the ABM Treaty. In Congressman Lee H. Hamilton's words this interpretation 'shakes up the world, our allies and Congress'.[19]

The 'new interpretation' has two versions. Both assert that the full-scale development and testing of ABM systems, including those that are space-based or use 'exotic technologies' or 'new physical concepts', are 'approved and sanctioned' by the Treaty. As to the deployment of such systems, according to the first version of the 'new interpretation' used by the White House and the Department of State, deployment is possible 'subject to discussion and agreement on specific limitation'.[20] In the meantime the Department of Defense adheres to the second version of the 'new interpretation' (in point of fact the fourth interpretation of the Treaty) and holds that the USA has the right even to deploy such a system without any limitations and subject to no preliminary agreement with the Soviet Union.[21]

Abraham D. Sofaer, Legal Adviser to the Department of State, in his statement in Congress in October 1985 asserted that the new interpretation 'is the Administration position on the meaning of the Treaty'.[22] As late as March 1984 the President in his report to the Congress wrote that the development, testing and deployment of space-based ABM systems, or components of such systems applying directed-energy technology, were prohibited by the Treaty. But already in October 1985 the US Administration, having made an about face, interpreted the Treaty in a directly opposite sense.

What is the rationale behind the 'broad' interpretation of the Treaty?

The main argument in support of the claim is the assertion that the Treaty in its totality covers only ABM systems and components that existed at the moment of its signing; as to components, the treaty is applicable only to those that are directly listed in Article II: namely, ABM interceptor missiles, launchers and radars.

The argumentation deliberately leaves out of the picture the first definitive part of Article II that reads: 'For the purpose of this Treaty an ABM system is a system to counter strategic ballistic missiles or their elements in flight trajectory...'. Secondly, undue stress is being laid upon the latter part of the phrase 'currently consisting of: (a) ABM interceptor missiles..., (b) ABM launchers..., (c) ABM radars...'. This closing part of the phrase is construed as proof that the given listing of components should be understood not in a functional sense, but as a limiting and exhaustive enumeration of all the ABM components falling under the Treaty provisions.

Common sense would not allow one to assume that states when entering upon a treaty of unlimited duration had the intention to perpetuate a limitation of ABM components that would cover only those types that already existed on the day the treaty was signed. The word 'currently' was inserted in the text of Article II precisely to stress the point that the ABM components listed further in the text did not exhaust *ad infinitum* that list. If the opposite were true, the word 'currently' would have been superfluous in the text. The word 'currently' stresses that in the future the composition of the ABM components might be different.[23]

Finally, any analysis of the Treaty should not overlook the fundamental provision formulated in Article I: each party undertakes 'not to deploy ABM systems for the defense of the territory of its country and not to provide a base for such a defense'.

Nevertheless the authors of the broad interpretation, pretending that everything outlined above is irrelevant to the point at issue, cite one of the agreed statements appended to the Treaty—the so-called 'D' statement. It states that in order to insure the fulfilment of the obligation not to deploy ABM systems and their components, except as provided in Article II of the Treaty, parties agree that in the event ABM systems based on other physical principles and including components capable of substituting for ABM interceptor missiles, ABM launchers or ABM radars are created in the future, specific limitations on such systems and their components would be subject to discussion in accordance with Article XIII and agreement in accordance with Article XIV of the Treaty.

The above statement is interpreted as permitting an unlimited creation and very nearly the deployment of ABM systems based on new physical principles (lasers, particle beams and others). Such an interpretation of the Treaty is diametrically opposed to the initial meaning attached to it. In point of fact the statement is ancillary to Article III of the Treaty, which provides for agreed-upon limited deployment areas for fixed ABM systems for each of the parties, and allows for the possibility of the appearance of ABM means based on new physical principles subject to the deployment restrictions of Article III and the ban on systems other than fixed, land-based.

Besides, the deployment of new fixed land-based means in areas allowed by the Treaty may be effected only after appropriate consultations by the parties on specific limitations and agreed-upon amendments to the Treaty, as provided for by its Articles XIII and XIV. Or, to put it otherwise, the intention behind statement D is the strengthening of the provisions of the Treaty banning the deployment of any large-scale ABM systems, and, at any rate, not the repeal of the prohibitions provided for by Articles I and V in respect to ABM systems and components based on new physical principles.

The Soviet side stated that the 'new' interpretation was an act of 'deliberate deception'.[24] The impermissibility of unilateral arbitrary interpretation of the Treaty was referred to by E.A. Shevardnadze, Soviet Minister of Foreign Affairs, and by Marshal S.L. Sokolov, Soviet Minister of Defence.[25]

The inadmissibility of the so-called 'broad' interpretation has also been stressed by US public figures who have been involved in negotiating the ABM Treaty, including Gerard C. Smith, Head of the US Delegation at the talks (see paper 2), John B. Rhinelander, Legal Counsel of that body, and Mr. R. Earle, former Director of the US Arms Control and Disarmament Agency.

Soviet and US lawyers jointly discussed this problem in March 1986. They arrived at the following conclusion:

Article II establishes a functional definition of ABM systems, which covers any system to counter strategic ballistic missiles or their elements in flight trajectory. Thus, the prohibition of Article V against the development and testing of space-based, sea-based, air-based or mobile land-based systems and components applies to systems using novel physical principles. This was the understanding of those who negotiated the Treaty, some of whom were at our session and of the bodies in each country that participated in its ratification. Indeed, the basic purpose of the Treaty noted above would be frustrated if development and testing of futuristic systems were freely permitted'.[26]

Four prominent US lawyers and former high-ranking US government officials made public their joint statement on the 'new' interpretation of the ABM Treaty. The statement was signed by Erwin Griswold, former Solicitor General and Harvard Law School Dean; Shirley Hufstebler, former US Court of Appeals Judge and Secretary of Education; Elliot Richardson, former US Attorney General and

Secretary of Defense; and Cyrus Vance, former Secretary of State and Deputy Secretary of Defense.[27]

Stressing that the ABM Treaty is 'the most significant treaty limiting the arms race between the United States and the Soviet Union' the signatories expressed their concern not only over the consequences of the actions performed by the US Administration, but also over the procedural aspects of such pursuits. They stated that the ABM Treaty, like any other treaty ratified by Congress, is legally recognized as the 'supreme law of the land'. When the Senate approved the Treaty (by a vote of 88 to 2), it acted on the understanding that the Treaty banned the development and testing of all space-based ABM systems and components, including 'exotic systems' based on novel physical principles. 'The Reagan's Administration's new interpretation', the authors go on to say, 'would amend the Treaty beyond recognition by eliminating a major constraint that has uniformly been regarded as an essential part of the Treaty since its ratification'. In conclusion the statement reads: 'Unilateral repudiation of the settled interpretation of this Treaty in order to facilitate new weapon schemes undermines the rule of law on which the international system is based'. The conclusions reached by US lawyers of such high standing are explicit in themselves and do not call for any further comment.

V. 'Practical policy'

A week had hardly passed after the 'new' interpretation was announced when sharp objections in the world press and in the Congress, as well as serious concern voiced by US allies, compelled the Administration to alter its posture on the issue. The Secretary of State, followed by other US officials, announced that at the current stage of the fulfilment of the SDI programme the US would, as a matter of 'practical policy', abide by its earlier interpretation of the Treaty, although, in their view, a 'broader' interpretation was legally justified.[28]

Paul Nitze, special adviser to the President and the Secretary of State for arms control matters, described the new posture of the US Administration in this way:

It is our view, based on our recent analysis of the treaty text and all of the accompanying records, that a broader interpretation of our authority than that which we have applied to restrict our SDI research program is fully justified. This is, however, a moot point. Our SDI research program has been structured, and, for solid reasons, will continue to be conducted in accordance with a restrictive interpretation of the treaty's obligations.[29]

This so-called practical policy of the US Administration as to its obligations under an international treaty calls for clarification on at least two points: in accordance with which restrictive interpretation of the Treaty will the SDI programme be implemented and how long will the Administration conduct such a practical policy? While these questions remain unanswered, the Treaty in no small measure loses its capacity as an instrument aimed at governing international relations in time of peace and at introducing elements of predictability into these relations.

In the statement quoted above Paul Nitze speaks of a restrictive interpretation, a fact indicative, evidently, of the presence of several other interpretations. As has been demonstrated above, not only the broad interpretation, but also the one that can be drawn from the documents of the Department of Defense and from a number of statements (made after the announcement of the SDI programme but before the appearance of the 'new interpretation') differ substantially from the understanding of the obligations under the Treaty that existed during its ratification period and for the 10 years that followed (in particular, as far as the admissibility of testing is concerned).

This point was also stressed by the Office of Technology Assessment of the US Congress: 'Both of the Administration positions [in April 1985 and October 1985] are at odds with the more restrictive pre-SDI interpretations of the Treaty'.[30]

In the absence of an answer to the second question posed above the US position on the ABM Treaty remains unpredictable. The record of the hearing in the House Foreign Affairs Committee reveals that an attempt to get the Administration to answer the question was futile.[31] This indicates the Administration's unwillingness to restrict itself to any obligations as to the duration of its practical policy; thereby it implies that it reserves the right to interpret in broader terms its authority under the Treaty and to do so at any time that, in its own judgement, best suits its intentions. All this is, undoubtedly, a fresh departure in the theory and practice of international treaty interpretation.

So, under the circumstances, where do the 'strict' and 'uniform' observance of the Treaty by the parties and the impermissibility of a 'double standard'—matters, purportedly, of such primary concern to the US Administration—come into the picture?

VI. Out of the international maze

Let us go back now and take up the reactions of Washington to the Soviet proposal made at Reykjavik on the prohibition of all space-based ballistic missile defence testing with the exception of research and testing conducted in laboratories. The various versions of this reaction may be summed up as follows: such a prohibition would: *(a)* be at variance with what the parties agreed when the Treaty came up for signing; *(b)* impose upon the SDI programme even stricter limitations than those existing at present under the ABM Treaty; and *(c)* render the ABM Treaty even more restrictive than under the 'narrow' interpretation. On the other hand, the US President insisted at Reykjavik on the right to carry on research and development 'as well as tests authorized by the ABM Treaty'.[32]

If the US side is rejecting the Soviet proposals and claiming that the tests are 'authorized' by the ABM Treaty on the basis of the so-called broad interpretation, any additional legal arguments against such a stand are useless. The absurdity of such an interpretation and its illegality in terms of international law and the national law of the United States has manifestly been proved by numerous prominent experts in the field of jurisprudence and politics (see, for instance, the statement quoted above by four well-known US lawyers and statesmen). In relation to such an interpretation of the Treaty the Soviet proposal might really be construed as restrictive. The US claim that tests are authorized by the Treaty turns into the 'right' to conduct any tests in outer space prior to the deployment of the system. This has nothing in common either with what the parties were negotiating while drafting the Treaty or with its interpretation during its ratification.

If, however, the US statesmen are guided in their arguments by the 'narrow' interpretation (and at least some of them assert that this is the case), the question arises what 'narrow' interpretation they have in mind: the one made public by the Administration during the ratification of the Treaty and reiterated up to 1984, or some other interpretation never explicitly formulated by the Administration? This is not an idle question. In the first case the Administration declared clearly to the public and the Congress that any testing of space-based ABM systems and components out of laboratory bounds was banned (this is exactly what was proposed by the Soviet side at Reykjavik as a common understanding of the obligation under the Treaty). In the second case, as shown above, the Administration, without clearly formulating its interpretation, introduces concepts non-existent in the Treaty such as 'subcomponents', 'devices' and 'technologies', the space-testing (or demonstration) of which has allegedly not been prohibited by the Treaty. This second 'narrow' interpretation is at odds with the Soviet proposal to confirm the prohibition on testing of all space-based elements of ABM defences in outer space.

If the United States really intends to secure strict adherence to the Treaty (and this intention has been declared by both parties to it), one can get out of this maze of interpretation, 'practical policies', deliberate and intentional reticence and ambiguity only by a specific accord stipulating which ABM activities are admissible under the Treaty and which are not. High-level talks on the issue were suggested by the Soviet Union during the meeting of the Soviet Foreign Minister with the US Secretary of State in Vienna in November 1986. The settlement of these problems in conjunction with an obligation not to withdraw from the Treaty during an agreed-upon term and to observe strictly all of its provisions would be practical proof of the intention of its signatories to strengthen the regime envisaged by the ABM Treaty, to prevent its erosion and, in the final analysis, the evolution of this accord into what is no more than a 'dead letter'.

Notes and References

1 Vladlen Vereshchetin is a professor of International Law and Deputy Director of the Institute of State and Law of the Soviet Academy of Sciences.

2 This is a quotation from the address by M.S. Gorbachev, General Secretary of the CC CPSU, broadcast by Soviet TV on 14 Oct.1986. See *Pravda*, 15 Oct.1986.

3 See, among other sources, the article 'Geneva: what does the first round of talks point to?', *Pravda*, 27 May 1985.

4 See, 'E.A. Sheverdnadze, Minister of Foreign Affairs of the USSR meets the press' (10 Nov.1986), *Pravda*, 11 Nov. 1986.

5 Sheverdnadze, E.A., Speech at the Conference on Security and Co-operation in Europe in Vienna on 5 Nov. 1986; *Pravda*, 6 Nov.1986.

6 *Pravda*, 14 Oct. 1986.

7 *ABM Treaty Interpretation Dispute*, Hearing before the Sub-committee on Arms Control, International Security and Science of the Committee on Foreign Affairs, House of Representatives, Ninety-Ninth Congress, First Session, 22

Oct.1985 (US Government Printing Office: Washington, DC, 1986), p. 273. See also: Longstreth, T.K., Pike, J.E. and Rhinelander, J.B., *'The Impact of US and Soviet Ballistic Missile Defense Programs on the ABM Treaty*, National Campaign to Save the ABM Treaty, Third Edition, Mar. 1985, p. 26.

8   See, among other sources, the Statement by M.S. Gorbachev, General Secretary of the CC CPSU, *Pravda*, 16 Jan.1986.

9   See: 'Memo From Charles Geliner, Congressional Research Service, Regarding Future Systems in Senate Hearings on the ABM Treaty', in: *ABM Treaty Interpretation Dispute* (note 7), pp. 139-44 and p. 106. It is noteworthy that Richard Perle, one of the architects of the 'new interpretation', attended those hearings in the Senate: ibid., p. 143.

10   *ABM Treaty Interpretation Dispute* (note 7), p. 354.

11   *Fiscal Year Arms Control Impact Statement*, Statements Submitted to the Congress by the President Pursuant to section 36 of the Arms Control and Disarmament Act, Senate Print 98-149, Mar. 1984. Worthy of attention is the fact that in a similar document submitted to Congress in 1985 that statement was dropped (for the first time in many years).

12   Numerous writers note with good reason that by analogy with US law such actions should be qualified as anticipatory breach of contract. See: Rhinelander, J.B., 'Implications of US and Soviet BMD programmes for the ABM Treaty', ed. B. Jasani, SIPRI, *Space Weapons and International Security* (Oxford University Press: Oxford, 1987); Doyle, F., 'Star Wars vs. International Law: the force will be against us', Doc. no. 3843 T, 1985, pp. 4, 6.

13   The theoretical foundation for such an approach can, evidently, be traced to the doctrine of the 'functionalists' who hold that the substance of a treaty can be changed through interpretation to make it suit the changing circumstances. It is appropriate to recall here that at the Vienna Conference on the Law of the Treaties the doctrine was upheld by the US Delegation, but when the respective amendment was put to a vote it was overwhelmingly rejected (by 66 'nays' and 8 'ays' with 10 abstentions). See: Talalaev, A.N., *International Treaties Law* (in Russian), Moscow (Mezhdunarodnye Otnoshenija Publishers: Moscow, 1985), pp. 79-80.

14   Department of Defense, *Report to the Congress on the Strategic Defense Initiative'*, 1985.

15   Department of Defense (note 14), p. B-1.

16   See: Sherr, A., 'The language of arms control', *Bulletin of the Atomic Scientists*, Nov. 1985.

17   Department of Defense (note 14).

18   *The Impact of US and Soviet Ballistic Missile Defense Programs on the ABM Treaty* (note 7), pp. 15, 29.

19   *ABM Treaty Interpretation Dispute* (note 7), p. 41.

20 *ABM Treaty Interpretation Dispute* (note 7), pp. 44, 55 and 153.

21 *ABM Treaty Interpretation Dispute* (note 7), pp. 48, 154.

22 *ABM Treaty Interpretation Dispute* (note 7), p. 49.

23 John Rhinelander writes: 'The phrase "currently consisting of" preceded by a comma, was included in the text to make clear that the definition is *functional*. The submittal letter from the Secretary of State Rogers to President Nixon dated June 1972 says so explicitly'; *ABM Treaty Interpretation Dispute* (note 7), p. 355. Further on the document testifies that the words 'currently consisting of' were inserted in the Treaty on the initiative of the US Delegation; ibid., p. 71.

24 See the article by Marshal S.F. Akhromeev, Chief of the General Staff of the Armed Forces of the USSR, 'Washington's allegations versus hard facts', *Pravda*, 19 Oct. 1985.

25 See the text of the contribution of E.A. Sheverdnadze to the general debate at the 40th Session of the UN General Assembly: *Pravda,* 25 Oct. 1985; also an article by Marshal S.L. Sokolov, Minister of Defense of the USSR, 'To preserve what has been achieved in strategic arms limitations', *Pravda*, 6 Nov. 1985; answers given to the press at a press conference in Moscow by G.M. Kornienko, First Deputy Foreign Minister of the USSR, and by Marshal S.F. Akhromeev, Chief of the General Staff of the Armed Forces, *Pravda*, 23 Oct. 1985.

26 Joint papers from the Fourth Conference of the Lawyers Alliance for Nuclear Arms Control and the Association of Soviet Lawyers, 24-31 Mar. 1986, p. 4.

27 *ABM Treaty Interpretation Dispute* (note 7), pp. 157-158.

28 *Washington Post*, 15 Oct.1985.

29 *ABM Treaty Interpretation Dispute* (note 7), p. 3.

30 *ABM Treaty Interpretation Dispute* (note 7), p. 133. The only definition of the so-called 'narrow' interpretation we managed to find was that given by A. Sofaer, Legal Adviser of the US Department of State: '...restricted interpretation, namely, research into, but not development or testing of, systems or components based on future technology and capable of substituting for ABM interceptors, launchers, or radars'—*ABM Treaty Interpretation Dispute* (note 7), p. 8.

31 *ABM Treaty Interpretation Dispute* (note 7), p.48.

32 Quoted from George Shultz's speech in the National Press Club on 17 Oct. 1986.

Paper 7. The ABM Treaty: verification and compliance issues

Richard N. Haass[1]
*Harvard University, John F. Kennedy School of Government, 79
John F. Kennedy Street, Cambridge, Mass. 02138, USA*

I. Introduction

The 1972 Treaty on the Limitation of Anti-Ballistic Missile Systems is widely hailed as one of arms control's cardinal accomplishments. The Treaty's severe constraints on US and Soviet deployment of anti-ballistic missile systems have strengthened a key dimension of contemporary deterrence, namely a common vulnerability to ballistic missiles and their nuclear warheads.

Yet this same treaty is increasingly the object of unprecedented challenge from a number of quarters. Part of this challenge is technological in nature. When the ABM Treaty was signed in 1972, both the actual and potential capacity of anti-ballistic missiles systems appeared modest; indeed, it was not so much the Treaty that made possible a world almost totally without ballistic missile defences as the absence of viable or envisioned defences that made possible a treaty. As a consequence of basic and applied research efforts in both the United States and the Soviet Union, however, a new generation of technologies that could pose a threat to ballistic missiles and their cargo is either beginning to appear or is imaginable. Much of the true potential of these new technologies could only be determined (not to mention exploited) if the Treaty were amended or discarded.

Political pressure is also taking its toll on the ABM Treaty. Individuals at various points along the political spectrum are rejecting contemporary deterrence and its foundation of mutual assured destruction. For some, the alternative is the abolition of nuclear weapons; for others, the alternative is defence. Moral reasoning aside, the perspective that the ABM Treaty is of questionable value has received added impetus from arms control's failure to constrain the quantitative growth and qualitative improvement of nuclear weapons. The situation has come about in which retaliatory capabilities can be considered suspect. Indeed, it is instructive to recall that just before the Treaty was signed, the US delegation notified the Soviets that 'If an agreement providing for more complete offensive limitations were not achieved within five years, US supreme interests could be jeopardized. Should that occur, it would constitute a basis for withdrawal from the ABM Treaty'.[2]

Political developments of another sort—in this case, intrinsic to the post-1972 history of the ABM Treaty—have also undermined support for the Treaty's remaining in force. Mutual US and Soviet charges of non-compliance with the Treaty have created an atmosphere of mistrust and there is concern that the Treaty's impact is uneven owing to the fact that one or the other party is acting in a manner inconsistent with its provisions. It is to this last set of political developments, those concerning verification and compliance, that this chapter is addressed.

II. The background to the debate

Before proceeding any further, some definitions are in order. 'Monitoring' refers to collecting and processing information about the activities of the other party. 'Verification' refers to the matching of information collected through monitoring against the provisions of an agreement to determine the degree of 'compliance'. Thus, whereas monitoring is solely a technical endeavour, verification is both technical and political in nature. Compliance in turn is principally a political judgement. 'Violations' refer to actions taken contrary to specific provisions of an agreement; that is, violations of the letter of the text. 'Circumvention', sometimes stated as 'violation of the spirit' or 'violation of the object and purpose' of a pact, refers most often to exploitation of imprecise treaty language, loopholes, omissions and/or ambiguities. While only violations have legal standing, both violations and circumventions possess political and possibly military significance.

At the time of the ABM Treaty's negotiation and signing, concerns over verification and compliance did not figure prominently in the public debate. They did, however, affect the terms of the Treaty itself. Article XII of the ABM Treaty explicitly endorsed not only the central role of national technical means of verification (NTM) to discern compliance but also precluded both parties from interfering with the NTM of the other or deliberately taking concealment measures that would impede verification. The ABM Treaty also established the Standing Consultative Commission (SCC) to, among other things, 'consider questions concerning compliance with the obligations assumed and with related situations which may be considered ambiguous'.[3]

Despite this effort to promote verification and minimize compliance concerns, the ABM Treaty has become embroiled in controversy. Indeed, although compliance concerns can be said to affect virtually all aspects of arms control, nowhere has the impact been as great as in the area of ballistic missile defence. This is not surprising. Violations of the 1972 Treaty could have an impact on the strategic balance more significant than violations of any of the agreements limiting offensive arms. Even extensive violations of either of the SALT pacts might, for example, increase a party's capabilities by a few per cent at most, not enough by itself to alter significantly the strategic balance; violations of the ABM accord could provide the basis of a defensive capability and thereby weaken the condition of mutual vulnerability upon which stability in the nuclear age has thus far been premised.

III. Current compliance disputes

A number of compliance issues have been raised by the United States.[4]

The first concerns Article VI(a) of the ABM Treaty, which prohibits the testing of air defence radars in 'an ABM mode. In the early years of the Treaty, the Soviets followed a practice of turning on their SA-5 radars located near or at ballistic missile test ranges at times of missile tests. Such 'concurrent operations' could help provide the SA-5 radar (an air defence radar) with ballistic missile tracking capabilities. The US challenged the Soviets on their behaviour, only to be told that the SA-5s were being used for air safety purposes (which are not prohibited) and not for ABM purposes disallowed under the Treaty. Subsequent discussions in the SCC to delimit concurrent use of air defence radars at missile test ranges narrowed but did not close the loophole.[5]

A second ABM radar-related compliance issue also emerged in the mid-1980s. The Treaty limited ABM testing to those ranges that existed at the time it was signed; testing could occur at new or additional ranges only by mutual agreement. In 1975, the United States observed that the Soviet Union was constructing an ABM radar on the Kamchatka Peninsula ballistic missile test range. When the United States challenged the USSR, the Soviets claimed that the facility had in fact existed at the time the Treaty was signed and therefore did not constitute a new facility requiring US approval.

The history, however, is more complex. During the negotiations, the United States asked the Soviet Union to confirm that the radar facility at Sary-Shagan was their single range, only to be told by the Soviets that national technical means of verification were sufficient to assess the situation. They never contradicted US statements that the USSR had only one test range at the time of the Treaty's signing.[6] The Soviet Union was guilty either of attempting to mislead the US side or of misrepresenting years later what had in fact existed at the time. In 1975 there was no way of proving what was the case, there being no annex to the Treaty that listed those facilities existing when it was signed.

A third radar-related compliance issue stems from Article V (1) and common understanding C of the Treaty and its prohibition of mobile radars, that is., those not of a 'permanent fixed type'. The specific Soviet systems at issue are the 'Flat Twin' and 'Pawn Shop' radars. Although the radars in question may not be truly mobile—the intent of the Treaty was to prevent the accumulation of radars that could be fielded quickly and thereby provide either party with the potential to break out rapidly of its limits—they are capable of being transported without too much difficulty and require little site preparation before becoming operational. They are as a result relatively movable. The components do not appear to violate the Treaty *per se* but come close to violating strict definitions of the letter; by so doing, the radars point up the difficulty in proscribing systems with specified characteristics when actual systems possess traits that do not necessarily conform to the categories established by negotiated agreements.

The most well known compliance issue, which has implications not only for the 1972 Treaty but indeed for all arms control, also concerns radars, in this instance the Soviet radar in Krasnoyarsk, Siberia. What is at issue again is a familiar dilemma for arms controllers: how to permit a given system in one set of circumstances while prohibiting it in another.

Limitations on large, phased-array radars (LPARs) constitute a key element of the restrictive regime created by the ABM Treaty. Such radars were and remain an essential component of an anti-ballistic missile system; because they take years to construct and are easily visible to satellites, limitations upon them are both necessary and easy to monitor. The problem stems from the desire of both parties to have such radars for other purposes: principally for early warning, space tracking and intelligence gathering.

In addition to limiting each party's ABM radars (initially to two sites at least 1300 kilometres apart; with the 1974 Protocol to only one site apiece) the Treaty permits the construction of additional LPARs for early-warning purposes so long as they are located on the periphery of the country's territory and oriented outwards. Moreover, the ABM Treaty does not prohibit the construction of LPARs anywhere in the country so long as they are for purposes other than ballistic missile defence or early warning, such as space tracking or intelligence collection.

In the case of the Krasnoyarsk radar, the United States has charged that both the siting and the orientation of the radar make it a poor candidate for space-related functions (which the Soviets say is the function) and that as a result it violates the Treaty's provisions governing early-warning radars. The alleged violation is of great concern to the United States both for political reasons—any violation indicates a lack of respect for and commitment to the arms control process—and for military ones, in that this radar is well located for such activities as ballistic missile warning, attack assessment and target acquisition and tracking.

The Krasnoyarsk case points out once more the danger of negotiating arms control agreements that allow the construction of systems possessing more than one application when one or more of the applications is restricted. The problem is compounded when the Treaty allows the deploying party to declare what the purpose of the system is; not surprisingly, the Soviets declare the purpose of the system to be one consistent with the Treaty's structures. By such reasoning the Soviet Union could seek to justify the deployment of unlimited numbers of LPARs anywhere within its borders so long as they could claim some space or intelligence-related function. Yet it was precisely the deployment of a large number of LPARs with their potential to provide the basis of a nation-wide, territorial ABM system that the Treaty was designed to prevent.

A fifth compliance issue raised by the United States concerns the alleged Soviet capacity for 'rapid reload' of its Galosh ABM interceptor test launchers kept at the Sary-Shagan missile test range. Article V (2) of the Treaty stipulates that neither party is to 'develop, test or deploy ABM launchers for launching more than one ABM interceptor missile at a time from each launcher' and in general not to develop the capacity for rapid reload. The Soviets have reportedly developed the capacity to reload a launcher in much less than a day. Given the uncertainty over just what capability the Soviets have and just what 'rapid' means, the United States has judged the situation to be ambiguous although of serious concern.

Another compliance matter raised by the United States centres not on radars but on a particular missile system. The system at issue is the SA-X-12, a modern surface-to-air missile system designed principally for air defence but possessing some capacity to engage ballistic missiles in flight. Since a wide range of missile systems possess at least some residual ABM capacity, the issue is not whether the SA-X-12 contains a capacity to engage ballistic missiles but how much. The United States has expressed its worry that the upgrading of systems such as the SA-X-12 provides the Soviet Union with some enhanced ABM capability, which in turn edges the USSR that much closer to a territorial defence.

The final compliance concern raised by the United States is to some extent the sum total of the others; namely, that 'the aggregate of the Soviet Union's ABM and ABM-related actions (e.g., radar construction, concurrent testing, SAM upgrade, ABM rapid reload and ABM mobility) suggests that the USSR may be preparing an ABM defence of its national territory'.[7] The charge received added impetus from the December 1986 announcement by Secretary of Defense Caspar Weinberger that the USSR is deploying an additional three LPARs.[8] Any such territorial network would directly contradict the Treaty's Article I, which prohibits the deployment of ABM systems for territorial defence as well as actions that would provide a base for such a defence. This prohibition forms the core of the Treaty's purpose; compliance with it is essential if the Treaty is to have meaning and if deterrence as we have come to know it can endure.

Taken as a whole, the US statements and charges add up to a concern that the Soviets are guilty of one clear violation of the Treaty (the Krasnoyarsk radar), a pattern of behaviour that indicates a regular disregard for its terms (or exploitation of its ambiguities) and that the USSR may be moving in the direction of a fundamental violation of (or breakout from) the agreement in the sense that Soviet activities viewed in their entirety constitute a territorial defence in the making.

For their part, the Soviets have consistently denied any wrongdoing or behaviour at variance with the ABM Treaty. 'The Soviet Union strictly complies with its international obligations...Violation of the letter and spirit of agreements reached, attempts to circumvent or undermine them by new military programs, refusal to ratify the agreements already signed—this, as is known, is not our policy'.[9] As regards the most important US allegation, that of the Krasnoyarsk radar, the Soviet government claims its location and orientation make it possible for the radar 'to fill a gap in the zone of monitoring by radars Soviet space objects and outer space in the Eastern part of the country'. The Soviets go on to add that all this will be able to be confirmed by the United States using NTM once the radar becomes operational.[10]

One possible sign that the Soviets are aware they have a problem on their hands of their own making is in their reaction to US objections to the Krasnoyarsk radar. It was reported that in October 1985 the Soviet Union offered in Geneva to halt its construction of the Krasnoyarsk facility in exchange for the forgoing of US plans to modernize its early warning radars at Thule, Greenland and Fylingdales, United Kingdom. The United States is said to have rejected the proposal on the grounds that the US radars, in existence when the 1972 Treaty was signed, were exempt ('grandfathered') from its constraints, and that any modernization was similarly outside the Treaty's purview.

At the same time, the Soviets have not been content with simply trying to rebut US charges; rather, they have gone on to levy charges of their own. As noted above, the Soviets have raised questions about the legality of the US upgrading of the Thule and Fylingdales radars, understandably so, given the debate within the United States and the United Kingdom as to whether the upgrading (which in effect transforms out-dated early-warning radars into modern ones of the phased-array type) violates the ABM Treaty's several clauses regarding radar construction.[11] In addition, the USSR has claimed that contrary to the Treaty's provisions, the United States is developing mobile ABM-related radars, testing an ICBM (the Minuteman I) to give it an ABM capacity, developing ABM interceptors with multiple warheads, laying the groundwork for a territorial defence, concealing ABM launcher silos, and establishing or upgrading radars with ABM-relevant capacities. The Soviet Union has also criticized the United States for violating the principle of resolving compliance disputes privately using the SCC.[12]

The United States has not responded to these charges with great seriousness, and the Soviets have not pursued them with much zeal. Instead, the principal Soviet concern with US behaviour relating to the ABM Treaty appears to be tied less to what the United States has already done than to what it is contemplating. In particular, the Soviets are concerned with the scope and potential of the Strategic Defense Initiative and its consequences both for the ABM Treaty and US ABM capabilities. In this concern, the Soviet Union seems in large part to be mirroring the US worry that the USSR is preparing for a territorial defence. Indeed, with the important exception of the Krasnoyarsk dispute, the compliance debate to come could well be more significant than the one we have seen thus far.

IV. The coming compliance debate

The coming or second phase of the compliance debate relating to the ABM Treaty is likely to be dominated by several issues stemming from its ambiguity regarding what types of activities are permitted. In each of these situations the debate has begun, and already it is apparent that the outcome of the debates will have major consequences not simply for the Treaty but for the entire political and strategic relationship between the United States and the Soviet Union.

Article V's first paragraph constitutes one of the Treaty's cornerstones. 'Each Party undertakes not to develop, test, or deploy ABM systems or components which are sea-based, air-based, space-based, or mobile land-based'. Not to be found anywhere, however, is a definition of what constitutes 'components'. As a result, each side is free to develop, test and deploy a wide range of technologies that it defines as falling below the threshold of 'component'. Similarly, the discretion is left to the parties themselves to determine the dividing line between permitted research and the other proscribed activities. Also possible is a debate over the legality of efforts to defend against 'tactical' ballistic missiles. The Treaty stipulates in Article II (1) that its purpose is to limit systems that might counter 'strategic ballistic missiles'—yet a good deal of technology could be developed and deployed to defend against tactical or non-strategic ballistic missiles that would have considerable relevance for countering the strategic variety.

There is even greater dispute over technologies not in existence in 1972 but increasingly available or at least imaginable. Two schools of thought have emerged: one that argues for a permissive interpretation of the 1972 Treaty, arguing that it did not significantly constrain the development of ABM systems based upon new futuristic technologies, and a second more restrictive school arguing that it did, that what was severely limited was not simply systems based on approaches then extant but any and all ABM systems.

For those arguing the former view, the key provision of the Treaty is agreed statement D, in which the parties agreed '...that in the event ABM systems based on other physical principles and including components capable of substituting for ABM interceptor missiles, ABM launchers, or ABM radars are created in the future, specific limitations on such systems and their components would be subject to discussion...'. It is argued that the mere existence of agreed statement D supports the thesis that the Treaty did not constrain ABM systems based upon exotic technologies; if it had, the statement would not have been necessary.[13]

The 'permissive interpretation' perspective gains further support from the negotiating record, namely the fact that the US side failed in gaining Soviet approval for broad definitions of the basic ABM system components (interceptor missiles, launchers, and radars) included in Article II (1). The Soviets specifically refused language proposed by the United States—'Each Party undertakes not to deploy ABM systems using devices other than ABM interceptor missiles, ABM launchers, or ABM radars to perform the functions of these components'—preferring that any discussion of 'futuristics' be kept outside the formal treaty and brought before the SCC if and when specific questions materialized.[14]

Those adhering to a more restrictive view of the Treaty maintain that the key prohibitions are to be found in Article V—'Each Party undertakes not to develop, test, or deploy ABM systems or components which are sea-based, air-based, space-based, or mobile land-based'—and that the definitions in Article II are illustrative rather then comprehensive.[15] The negotiating record is again cited to bolster the argument, in this case to demonstrate that the intent behind agreed statement D was simply to reinforce the ban on deployments (while maintaining the possibility) of fixed land-based systems based upon futuristics, something politically necessary given the position of the Joint Chiefs of Staff and legally necessary given the fact that this was the one form of ABM system explicitly permitted by the Treaty. Additional support for this interpretation can be garnered from the hearings and debate in the US Senate over ratification of the ABM Treaty.[16] The conclusion of the 'restrictive interpretation' school is that the Treaty permits development and testing but not deployment of fixed land-based futuristics and prohibits all but research of ABM systems of any other nature.

Not surprisingly, the Soviet Union is an advocate of the restrictive approach to the Treaty. The current Soviet view is that Article V covers both the then-existing technologies as well as futuristics, and that agreed statement D refers only to fixed land-based systems. Animating this Soviet position is the belief that the new US interpretation (announced in October 1985) is little more than a legal pretext for going ahead with the Strategic Defense Initiative.[17]

The intention here is not to settle the disagreement. It is, however, essential to recognize that the textual ambiguity and uncertain negotiating record provide the raw material for a compliance debate at some time in the future that could well dwarf its predecessors.[18]

V. The ABM Treaty—lessons and prospects

A good number of problems that have some to plague the ABM Treaty regime were predictable if not entirely avoidable. There is a good deal of ambiguity (or lack of specificity) in the Treaty text. The absence of clear definitions making clear just what constitutes a 'component', or the absence of a dividing line between 'research' and 'development' and 'testing', all but ensures compliance problems down the road.

Yet avoiding much less doing away with such problems is easier said than done. The Soviet Union tended to resist US attempts to reduce ambiguity, while in certain circumstances both sides were reluctant to specify certain activities for prohibition for fear of compromising sensitive intelligence sources and methods. Specificity brings with it another problem as well: much as is the case with a tax code that attempts to detail conditions under which tax must be paid only to create avenues for avoidance, so too can detailed accords create opportunities to 'design around' precise language. Perhaps most important, the process of negotiation often involves a degree of ambiguity if agreement is to be reached between two parties. Soviet resistance to US efforts at negotiating careful language affecting futuristic technologies has now come back to haunt the Treaty's implementation, as can be seen in the debate over agreed statement D.

The Treaty is also weakened by the disputes over systems that are permitted in one set of circumstances but not in another. Unless a system is banned altogether (or banned explicitly from operating at certain locales or in specified situations) a loophole is created by which a system with more than one potential purpose can be deployed even if one possible use is prohibited. Many of the disagreements over missiles with residual ABM capabilities, as well as the more important matter of the Krasnoyarsk radar, demonstrate this truth. So long as there are dual-purpose systems, and so long as parties are permitted discretion in their use, the potential exists for Treaty provisions to be violated or circumvented in a manner in which the other party to an agreement is left with little or no means to gain satisfaction as regards its claim of noncompliance.

The alternatives in such situations are several: to ban any system with certain potential applications either entirely or in specified circumstances; to permit specific deployments only as mutually agreed, thereby giving the other party an effective veto; to require the addition of truly functionally-related observable differences (FROD) characteristics that would limit a system's potential to be used for unwanted purposes, and/or to share intelligence data demonstrating the illegal application of a system once it is operational. As is obvious, none of these 'solutions' is without its own shortcomings. Also readily apparent is that it is unrealistic to expect the SCC to resolve ambiguities that years of negotiating could not.

The future of the ABM Treaty was also made less certain than it might have been owing to the almost total lack of explicit data bases. Both the US unhappiness over the number of Soviet ABM test ranges and the Soviet displeasure with the US upgrading of the Thule and Fylingdales radars might have been avoided if more complete and commonly accepted data bases had accompanied the initial Treaty. It should be recognized, however, that negotiating such annexes can be a difficult and prolonged exercise; the Mutual Force Reductions negotiations, fruitless after more than a decade of effort, offer an example of how data disputes can make progress on arms control all the more difficult.

What, then, can one predict about the ABM Treaty as it reaches its fifteenth birthday? Several distinct futures can be envisioned. One might be characterized by a gradual unraveling of the accord, to be brought about by a growing gap between the two parties over what constitutes compliance and growing incentives to determine or exploit the potential of emerging technologies. Each side could assume an increasingly broad or permissive interpretation of what the Treaty permits. Indeed, some in the United States would argue that this process has already begun owing to Soviet non-compliance; some in the USSR would claim that the United States has revealed its intention to proceed in this fashion as a result of the March 1983 address to the nation by President Reagan, establishing SDI, and the declaration of a new and more permissive interpretation of what the Treaty allows.

If the two countries do in fact move in this direction, a key consideration could be the process and style of change. Unraveling could occur quickly and with a good deal of unstructured unilateralism; alternatively, it could occur gradually and with a significant element of tacit co-operation. If the latter is the case, we may see the emergence of an

ABM 'regime', less formal than the original treaty but an attempt all the same to cope with technologies and strains not foreseen in 1972.[19]

The above description of possible paths or futures suggests the two principal policy alternatives. One is renegotiation, an explicit negotiation leading to a new accord or a protocol to the existing one in which new specificity is added. Such an effort might pass judgement on existing matters of dispute (particularly radars) as well as indicate what kinds of research, development and testing are permitted. It might also alter the withdrawal provision of the Treaty to provide for a notification period longer than the current six months. Still another possibility would be to give the Treaty a minimum fixed length, to a certain date, so that at least until that date the parties could be more certain of the environment they would face.

The other principal policy option is abrogation, either formally through one or both parties' exercising their right under the Treaty to withdraw with six months prior notice, or informally through one party's declaring the Treaty to be null and void and itself no longer bound owing to the behaviour of the other. This is the extreme form of unraveling, although it too could lead to the creation of an informal regime or even a new treaty.

For better or worse, the path that seems least likely is to turn the clock back to 1972 and live under a highly restrictive formal treaty. Changing politics and technology have brought about new pressures on the ABM Treaty. The challenge lies in the question of where to adapt the Treaty regime and where to resist; what is certain is that past, present and future compliance disputes will complicate this challenge to an extent unanticipated by those who negotiated this Treaty a decade and a half ago.

Notes and references

1   Richard N. Haass is lecturer in public policy and senior research associate at Harvard University's John F. Kennedy School of Government. A former official in the US Departments of State and Defense, he is the author of *Congressional Power: Implications for American Security Policy* (IISS: London, 1979) and co-editor of *Superpower Arms Control: Setting the Record Straight* (Ballinger Publishing Co.: Cambridge, Ma., 1987).

2   In Unilateral Statement A of 9 May 1972.

3 The text of the 'Treaty Between the United States of America and the Union of Soviet Socialist Republics on the Limitation of Anti-Ballistic Missile Systems' can be found in many anthologies. The one used here is US Arms Control and Disarmament Agency, *Arms Control and Disarmament Agreements: Texts and Histories of Negotiations* (US Government Printing Office: Washington, DC, 1982), pp. 139-47. This includes not only the basic treaty but also all agreed statements, common understandings and unilateral statements.

4 For the essential background to US compliance concerns, see President Reagan's reports to the Congress of 23 Jan. 1984, 1 Feb. 1985 and 23 Dec. 1985; the Oct. 1984 report by the General Advisory Committee on Arms Control and Disarmament, *A Quarter Century of Soviet Compliance Practices under Arms Control Commitments: 1958-1983;* and the Feb. 1986 report of the US Arms Control and Disarmament Agency entitled *Soviet Noncompliance*. For comprehensive treatments of the compliance debate, see Schear, J.A., 'Arms control treaty compliance' in *International Security*, vol. 10, no. 2 (Fall 1985); and Voas, J., 'The arms control compliance debate' in *Survival*, vol. 28, no. 1, (Jan.-Feb. 1986).

5 A June 1985 'common understanding' reached in the SCC established a mechanism under which either party could challenge the other on alleged illegal concurrent testing of ABM radars in an air defence mode. The understanding does not prohibit the existence of air defence radars at ABM test ranges or even their operation during missile tests so long as the radar is being used to promote range air safety, i.e., to prevent aircraft in the vicinity from straying into danger amid an ABM test. Thus, at least in principle, banned ABM-related concurrent testing could still occur under the cover of permitted air defence activities. For a more generous interpretation of the significance of this understanding, see Jeffrey Smith, R., 'Arms agreement breathes new life into SCC' in *Science*, 9 Aug. 1985.

6 In common understanding B, the United States asserted that there were two US test ranges (at White Sands, New Mexico and Kwajalein Atoll) and one Soviet range (at Sary-Shagan in Kazakhstan). The Soviet Union stated in response that 'there was a common understanding on what ABM test ranges were...and that national means permitted identifying current test ranges'.

7 US Arms Control and Disarmament Agency, *Soviet Noncompliance* (US Government Printing Office: Washington, DC, 1986), p. 6. To my knowledge, the most complete official US description of Soviet ballistic missile defence efforts that is available publicly was given by Robert M. Gates, the Deputy Director of Central Intelligence, to the World Affairs Council of Northern California on 25 Nov. 1986.

8 See the remarks of Secretary of Defense Caspar W. Weinberger to the American Legislative Exchange Council, 11 Dec. 1986, Washington DC (Department of Defense News Release 608-86). The existence of the new radars was actually made public by Gates's speech of 25 Nov. (see note 7 above) but it was not until Weinberger's address that the announcement gained much notice. See also Skidmore, D., 'US says Soviet facility could violate ABM pact' in *The Boston*

Globe, 12 Dec. 1986, p. 3. The Defense Department's concerns may not be universally shared within the US Government. See Gordon, M.R., 'US is debating role of 3 new Soviet radars' in the *New York Times* , 19 Dec. 1986, p. A11.

9
This is an excerpt from a response of the Soviet Government of 23 April 1985 to Les Aspin and other US Congressmen who had written to Gorbachev on 20 Mar. 1985 regarding compliance issues. The response was unsigned and hand-delivered by the Soviet embassy in Washington.

10
Hand-delivered response by the Soviet Government (note 9). For additional, more detailed Soviet defences of their behaviour, see the Tass statement, 'USSR hasn't broken arms pacts' in *Pravda Press,* vol. 36, no. 42, (14 Nov. 1984), pp. 4-5; and Sokolov, S.L.,'Preserve what has been achieved in strategic arms limitation' in *Pravda,* 6 Nov. 1985, reprinted in *The Current Digest of the Soviet Press,* vol. 37, no. 45 (4 Dec. 1985), pp. 24-25. Sokolov was the Soviet Minister of Defense.

11
See Gordon, M.R., 'US pursues plan for new radars despite fears of treaty violation' in the *New York Times* , 28 Dec. 1986, pp. A1, A6. The key language concerns Article VI (b), in which each party undertakes 'not to deploy *in the future* radars for early warning of strategic ballistic missile attack except at locations along the periphery of its national territory and oriented outwards' (emphasis added). The Reagan Administration argues that because warning radars existed at these sites all modernizations are permitted, given the Treaty's concern with future radars; critics of the policy charge that the modernization is so far-reaching in scope as to constitute the deployment of a new radar and hence violates Article VI (b).

12
For Soviet criticism of US compliance behaviour see 'The US violates its international commitments' in *Pravda,* 30 Jan. 1984, reprinted in English in *The Current Digest of the Soviet Press,* vol. 36, no. 4, (22 Feb. 1984), pp. 4-5; and 'Don't sabotage commitments, observe them' in *Pravda,* 9 Feb. 1985, reprinted in *The Current Digest of the Soviet Press,* vol. 37, no. 6, (6 Mar. 1985), pp. 7-8.

13
See the statement by State Department Legal Advisor Abraham D. Sofaer in *ABM Treaty Interpretation Dispute,* Hearing before the Subcommittee on Arms Control, International Security and Science of the Committee on Foreign Affairs, House of Representatives (US Government Printing Office: Washington, DC, 1986), pp. 9-18. This volume of testimony and documents is indispensable for anyone wishing to understand the debate over agreed statement D and the entire 'futuristics' debate concerning the ABM Treaty.

14
The negotiating record remains classified. However, see the important discussion of the futuristics matter in Smith, G., *Doubletalk: The Story of the First Strategic Arms Limitation Talks* (Doubleday: New York, 1980), especially pp. 263-65, 343-44; see also paper 2 in this volume.

15
See, for example, the testimonies of Gerard Smith, Ralph Earle and John Rhinelander in *ABM Treaty Interpretation Dispute.* (note 13). Also see the

letter of 1 Dec. 1986 from Senator Carl Levin to Secretary of State George Shultz in which the Senator charges the State Department Legal Advisor with providing Congress 'an incomplete and misleading analysis of the (negotiating) record' and with conducting a 'fatally flawed' review in reaching the new interpretation.

16 Ibid., p. 39.

17 See Akhromeyev, S., 'Washington's claims and the real facts', in *Pravda*, 19 Oct.1985), reprinted in English in *ABM Treaty Interpretation Dispute* (note 13), pp. 305-9.

18 The future may be arriving sooner than expected if those in the US Government favouring 'early deployment' of SDI systems prevail. For background see Sanger, D.E., 'Many experts doubt "Star Wars" could be effective by the mid-90's' in the *New York Times*, 11 Feb. 1987, pp. A-1, B-13; and Gordon, M.R., 'Arms debate now centers on ABM pact', in the *New York Times*, 17 Feb. 1987, pp. A-1, A-6.

19 There is a substantial literature on 'regimes'. 'Regimes can be defined as sets of implicit or explicit principles, norms, rules, and decision-making procedures around which actors' expectations converge in a given area of international relations'; Krasner, S., 'Structural causes and regional consequences: regimes as intervening variables' in S. Krasner (ed.), *International Regimes* (Ithaca: Cornell University Press, 1983), p. 2. In the same volume, Robert Jervis sets out preconditions for establishing a regime, the key one being a shared desire by the parties involved to live in a more regulated environment. (See his chapter 'Security regimes', especially p. 176.)

Part V. The international dimension of BMD systems

Paper 8. Ballistic missile defences into the next century—implications for the United Kingdom's strategic deterrent forces

Ronald Mason[1]
Chestnut Farm, Weedon, Ailsbury, Berkshire HP22 4NH, England

I. The broad background

It is almost four years since we knew of the Presidential challenge to render nuclear weapons 'impotent and obsolete'. Much has passed by since that challenge was laid down. A variety of political, economic and military issues have been raised, most of them finding different emphases in Western Europe than those which determined the debate in the United States.

The central politico-strategic issue in the debate has related to the question of whether the Strategic Defence Initiative, defined in one way or another, will enhance or diminish deterrence and international stability; a stability which must, in part at least, be attributed to the Western alliance's strategy of flexible response and, in a different but complementary way, to adherence to the provisions of the ABM Treaty. Related to the latter point are the possible linkages which connect deep cuts in strategic arms with the nature of reasons for deployment of strategic defences. Perceptions of linkages were sharply focused in the wake of Reykjavik and demonstrated a remarkable consensus in Western Europe that obsolescence of ballistic missiles or, more generally, of nuclear weapons was not for the foreseeable future. The Europeans have become insistent on contributing to any arguments concerned with moving beyond the research phase of SDI. The Federal Republic of Germany, Italy and the United Kingdom have established Memoranda of Understanding with the United States to facilitate co-operation on strategic and theatre defences; France is developing a robust policy for industry-industry collaboration. Without doubt an interest in access to advanced US technologies was a stimulus to co-operation. There was and indeed is a determination to understand whether there are new technologies which could not only be used in weapon systems that would in a general way contribute to NATO's defence but might also have important implications in the civil sector. For the two European nuclear weapon states, there must also be an interest in understanding strategic and theatre defence architectures, for it is only then that the balance between measures, countermeasures and counter-countermeasures can be sensibly assessed; where to use the Nitze phrase 'cost effectiveness at the margin' can be calculated; when the requirements and definition of the British and French strategic nuclear forces can evolve in response to a changing threat (strategic or quasi-strategic defences) environment.

It remains to be seen whether the Europeans will develop a significant programme of *technology* co-operation in which the particular technologies relate to strategic defences, as envisaged by President Reagan, rather than to those of a theatre defence initiative—the so-called intermediate options. A theatre defence initiative in Western Europe will emphasize mid-course and terminal-phase interception It will be based on a recognition that ballistic missiles are a small fraction of the total threat; that conventional and chemical payloads are or may become significant threats; but that there are some generic technologies and systems which will be common to strategic and theatre defence. In this respect there are some major policy decisions to be determined in connection with the deployment of tactical ballistic missile defences.

II. The UK deterrent

With that broad background in place we can now look at some of the specifics surrounding the modernization of the United Kingdom's strategic force. Government policies have been set out in three Open Government Documents—80/23, 82/1 and, very recently, 87/01. A *personal* retrospect on these papers and the background to them would highlight the following:

1. There is a commitment, on the part of the United Kingdom, to the importance of a 'second centre of decision' within the alliance, the existence of which must complicate the assessments of any potential adversaries and therefore contribute to the alliance's deterrent postures.

2. This would provide considerable insurance against the possibility that the Soviet Union would miscalculate the strategic commitment of the United States to the defence of Western Europe.

3. Any discussion of a European deterrent force, involving some form of co-operation between France and the United Kingdom, needs to consider the management of resources with more comprehensiveness than would ensue from a simple preoccupation with the desirability, or otherwise, of a joint development of missile programmes.

4. There appear to be significant differences in strategic doctrines between the United Kingdom and France.

5. The fundamental deterrent posture derives from an assured retaliatory capability, a capability which assesses, as carefully as one can for a period of 30 years or more, the emergence of countermeasures and a (national) ability to evolve and sustain a programme of counter-countermeasures.

6. An increased accuracy of delivery of re-entry vehicles permits deterrence to relate to specific interdiction against the strategic (political, military and industrial) assets of an adversary, rather than rely on the (less acceptable) strategic doctrines of World War II (one might add that it is also the correlation of accuracy with specificity of interdiction which may provide quasi-strategic value to conventional weapon systems).

7. The United Kingdom could not sustain a triad of strategic assets; a diad of strategic and theatre nuclear forces could constitute a valuable contribution to the alliance's doctrine of flexible response.

8. The survivability requirement for retaliatory capabilities advances the argument for an underwater platform (with the understanding that judgements have been made on the likely advancements in anti-submarine warfare). It also requires judgements on the relative sensitivity of ballistic and cruise missiles to defences, including 'breakouts' from the Anti-Ballistic Missile Treaty; developments in stealth technologies and systems; and developments in acquisition and discrimination technologies.

III. Examining defence concepts

The United Kingdom believes that, for the foreseeable future, a conservative interpretation of the ABM Treaty is appropriate. This is not a view which has its origins in a simple protectionist attitude—that breakdown of the ABM Treaty could have a profound effect on the viability of the British strategic force; rather, it is built from concerns which are reflected in the four points of agreement reached between the Prime Minister and the President at Camp David and in the paper by Sir

Geoffrey Howe presented at the Royal United Services Institution in 1985. A central issue concerns the transition, if it were to take place, from the research phase of a strategic defence initiative to the deployment of a comprehensive or partial defensive system. It is being currently suggested in the United States that there have been striking developments in certain technologies (largely sensors and guidance and control) which constitute *a priori* reasons for deployment of, for example, an advanced terminal intercept system. That view should not obscure the fact that there is an enormous gap between the solution of incorporating technologies in a survivable strategic system.

Figure 1 shows schematically how studies of the requirements and definition of an architecture—for strategic and theatre studies—are proceeding.

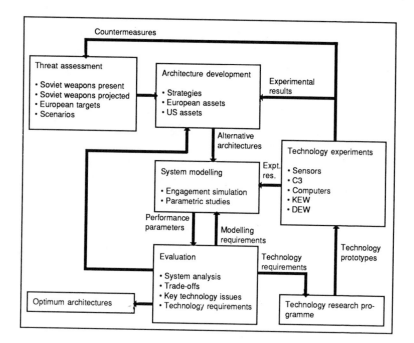

Figure 1. Architecture study structure

The figure illustrates:

(a) the classical interaction of 'the threat' against alliance assets; any analyst will recognize the uncertainties attached to threat evolution over

three to four decades, which may or may not be susceptible to rational sensitivity analysis;

(b) the need for a 'many on many' calculation, the methodology of which is in need of refinement; and

(c) that technologies *per se*, and their dynamics of development, do not, in any causal way, provide clear indicators on how cost-effective, survivable systems should be constructed and integrated.

The European nuclear weapon states must, in all prudence, maintain an understanding of how any theatre defence initiative *(vide supra)* should link with a CONUS (continental US) strategic defence initiative. For France and the United Kingdom, the space segment of any defence initiative is likely to be concentrated upon surveillance and communications capabilities and it is obviously in the interests of these two countries to have certain independent and survivable space assets; collaboration in space, perhaps with other Western European countries, would have a more constructive contribution to make to a European deterrent force, however that force be defined, than would follow from a joint missile development. The United Kingdom, with as sophisticated a background in countermeasures as exists in the alliance, can be expected to carry out research programmes in signature and discrimination technologies in order to protect its strategic investments—only verifiable agreements on strategic arms control and agreement on a new interpretation of the ABM Treaty will call for the significant adjustment of present plans for future strategic forces.

The most recent debate in Western Europe is removed from the immediate concern of the implications of SDI for national deterrent forces but, arguably, will affect policy options in the alliance in a more direct way.

IV. Analysis of the threat

The immediate technical debate in Europe, after the launch of SDI, was connected with the nuclear threat; the very least that should be said on behalf of SDI was that it brought into focus, for a wide audience, the total spectrum of threat. That range may be widened by increased accuracies of tactical missiles (missiles with ranges of under 1000 km) with advanced conventional and chemical payloads. Such conventional capabilities, if they were to exist, would have profound effect on the alliance's flexible response doctrine. Anti-tactical ballistic missile

(ATBM) defences, directed against conventional tactical ballistic missiles, can be thought of as integral to any evolving air defence system (whose targets would include cruise missiles, stand-off weapons and fixed-wing aircraft) and would not necessarily have great overlap with a defensive system directed against nuclear-tipped re-entry vehicles (a subset of a strategic defence initiative).

Critical analysis of the threat spectrum continues: when and how will the Soviet Union invest its tactical missiles with, say, the accuracy of a Pershing II system; what is that state of development of conventional sub-munitions and dispenser technologies; what criteria will determine the allocation of tactical missiles with conventional, chemical and nuclear payloads respectively; what perceptions will the Soviets have on arms control issues (verification matters related to type of payload and type of missile); and post-Reykjavik, the scale of the Soviet ballistic missile force.

When postulating possible defence systems, great care is therefore necessary in judging the threat spectrum to be defended against, for such a programme can vary enormously in sophistication and cost. Inevitably, assets to be protected are chosen restrictively and have tended to focus on airfield denial. For such targets the requirements for missiles are not small and can only be significantly reduced by chemical if not nuclear missile payloads. The conventional tactical ballistic missile threat must, in short, be seen in context as but part of the aggregate threat: useful, quick to apply, but limited. However, it remains true that a tactical ballistic missile strike against NATO's high-value targets could be damaging for in the final count it is penetration and defence suppression, prior to more massive attack by conventional carriers, which constitute a threat unmatched by presently deployed air defence systems.

The alliance must over the near-term future consider its response to the evolving conventional and chemical threat and must analyse the relative priorities and effectiveness of passive, active or offensive countermeasures. Dispersal, redundancy, mobility, hardening and other passive defence measures are all important in complicating target plans and there can be little doubt that such measures should enjoy some emphasis. On the question of active defences—which takes us closer to our central concern in this paper—the argument is that technology can now provide limited point defence against certain classes of ballistic missile threat, without recourse to nuclear kill mechanisms, and can give a realistic response to a wide range of air-breathing threats.

But in moving from a requirement of 'warhead kill' to 'mission kill', from limited area defence to genuine area defence, the West will have to consider costs and doctrines very carefully. Undoubtedly in so doing the cost-effectiveness of other options, such as counterforce, will enter the debate; and they may emerge as a particularly strong option in their compatibility with arms limitation negotiations.

V. Political and technical priorities

The limitations of the ABM Treaty with regard to defences against aircraft, non-ballistic missiles, non-strategic ballistic missiles and so on are well-known, as is the fact that transfer of ABM technology refers only to strategic systems. But Europe in general and the UK in particular set considerable store by the ABM Treaty, attach themselves to a conservative interpretation and believe that a full exo-atmospheric anti-tactical ballistic missile system deployed in Europe would undermine the spirit of the Treaty. The political (and technical) priority must be the devising of an effective endo-atmospheric system.

If that analysis is roughly correct, the implications for the planned United Kingdom strategic deterrent force can be thought through with some precision. The 'threat' to the European deterrent forces lies more with the bids to shift investment to more 'stable deterrents' such as cruise missiles. Whether those bids will continue to be comprehensive ones or will, more rationally, emerge as quantified proposals which seek deterrence at much lower levels of strategic nuclear weapons are the questions that will preoccupy thoughts over the next two years.

Notes and references

1   Sir Ronald Mason is a former scientific adviser to the British Defence Ministry.

Paper 9. The implications for France

Dominique Moïsi[1]
IFRI, 6, rue Ferrus, F-75683 Paris, France

I. A revisionist America?

In his first essay on the European diplomatic system at the time of the
the Congress of Vienna, 'a world restored', Henry Kissinger
differentiated between the *status quo* powers Great Britain and Austria
and the revisionist powers Russia and Prussia. In his mind such a
scheme applied equally well to the modern world and in particular to the
division between the revisionist Soviet Union and the pro *status quo*
United States.

To understand France's position in the wake of the debate on strategic defence started by the 'Star Wars' speech of President Reagan and
accelerated by the Rejkjavik summit of October 1986 and the subsequent resumption of negotiations between the Soviet Union and the
United States, one must realize that France is above all a *status quo*
power, keen on preserving her nuclear status and satisfied with her independent position within the Atlantic alliance. Therefore France can
react only with the utmost reservation when she sees the US turning
towards the adoption of 'revisionist' stances. This major reservation
cannot, however, be expressed fully given the fact that France has also
to take into account other European reactions, especially in the Federal
Republic of Germany, and the new factor introduced by a smiling and
dynamic leadership in the Kremlin.

II. France and SDI

Alice started to laugh. 'It's not necessary to try', she said. 'One cannot
believe impossible things'. 'You don't have much practice', said the
queen. 'At your age, I happened to believe up to six impossible things
before breakfast'. This quotation from Lewis Carroll, although not
taken from the recently published book by François Mitterrand
Reflections on French Foreign Policy,[2] aptly expresses the basic
thinking of the French President on SDI. However, incredulity is not
the only element of the debate.

In fact, looking at French attitudes towards SDI, one can see a contradiction, a tension between a clear diffidence as far as strategic concepts are concerned, and a mixture of temptation and fear with regard to the industrial implications of the project. This complex attitude can be found in the political opposition as well as in the majority—and, of course, today in a time of cohabitation one cannot distinguish between the opposition and the majority. To try to differentiate specifically between the right and the left on the matter, is simply false. It is true that that there are nuances between the two sides on SDI, but there are as many nuances in the conservative parties' support for SDI as there were nuances in the Socialist Party's diffidence toward the project. Raymond Barre, serious contender for the presidential election in 1988, has always demonstrated a clear diffidence towards the project, unlike the Prime Minister, Mr Chirac, who has alternated between phases of enthusiasm and restraint.

Against SDI

France's specific attitude stems from the fact that, unlike Japan and Germany, she has based her security on nuclear deterrence. In fact, it is very common in France to hear the comment that through SDI Germany and Japan could transcend a nuclear world from which they were excluded as a result of World War II and subsequent treaties. For France, it is just the reverse.

Unlike many people in the UK, Frenchmen do not believe in a special relationship with the United States that would make France a special case regarding technology-sharing with the Americans. France's attitude has been fuelled by the contradiction in the US presentation of the project. By transforming a rather classical debate about the comparative virtues of offensive and defensive nuclear weapons into a millenarian vision, the great communicator, President Reagan, knew what he was doing in his attempt to mobilize US public opinion. But that vision was bound to clash with the other more diffuse presentation of SDI that was later proposed to the Europeans, emphasizing the complementary nature of defensive deterrence and offensive deterrence.

The clash between a scheme emphasizing continuity and a vision discarding it was seized upon by the French, two years after the March 1983 speech of President Reagan, as a way to justify their diffidence *vis-à-vis* the project. For the French, SDI represents, above all, a major political and strategic risk. Any attempt to present SDI as a substitute security system that would replace today's world based on nuclear deterrence is most dangerous. At the time of the Euromissile quarrel, when the West was confronted with pacifist waves, France was basing its political strategy on the fact that deterrence had worked. The success of this joint battle has been the restoration of the legitimacy of the centrality of the nuclear component in the defence of Europe. This has been done in the face of the attacks of those who are using the social dimension of strategy. In the name of a millenarian, globalist and optimistic vision of history, how could the Americans abandon us in the midst of the battle when the Soviets are resuscitating the Euromissile question? That was seen as totally unacceptable for France. In fact, in their rejection of the US scheme, the French combined historical scepticism and scientific incredulity.

In Europe people feel that history is by essence tragic. The idea that weapons could successfully fight war above the heads of the citizens does not convince the Europeans. Equally, the challenges of science and nature are such that no one today can demonstrate that the system of special defence can work. France's scepticism on that matter is reinforced by the debate within the United States itself between pro- and anti-SDI experts.

Beyond that global political risk, there is a special strategic risk for France. SDI is the key element of a defence system mainly based on the threat of Soviet ICBMs. It is the culmination in the United States of a reflection on the window of vulnerability. But the SDI concept does not correspond to the diversity of threats facing even the United States, and it is less capable of addressing the threats to Europe. Within our continent, we are confronted with a range of threats, from conventional to nuclear, and the conventional imbalance is such that we believe there is no alternative to nuclear deterrence.

One can, therefore, understand the US position and its presentation of ideas as an attempt to satisfy four categories of public: the pacifists who are against nuclear weapons; the moralists and the humanists who are against MAD (mutual assured destruction); the unilateralists who visualize the specific protection of the United States from all threats; and the suprematists, those who want to restore the supremacy of the USA. But such groups are weak or do not exist in France, and therefore we cannot share that line of argument. (In particular the pacifist ecological group in France has been particularly discreet, partly out of cynicism towards nuclear matters and partly because these groups appear to be manipulated by a weakened communist party.[3]) There is a strong fear that future prospects will provoke, today, a Soviet reaction: an escalating arms race which will reduce the credibility of the French nuclear force. In the name of greater security tomorrow, Europe would find itself today in a more destabilizing environment—one that would be particularly damaging for France's independent security posture.

Moreover, the French position has been to reject the idea that a specific stance on SDI was a test, a criterion, with which to judge a good ally. Ultimately, France was a good ally before, it still is a good ally, and it does not have to show that by rallying to SDI.

French ambivalence

From the above it appears one would have to characterize France's attitude as sheer rejection based on refusal to accept the evolution of strategic concepts. Yet, France's attitude has been, from the start, much more ambivalent and this ambivalence has been reinforced lately. It stems from what is judged as realism, and from a mixture of temptation and apprehension *vis-à-vis* the technology and industrial dimensions of SDI.

Realism? SDI, as a research effort is going on. It transcends the personal choices of President Reagan. Its budget might be drastically cut by Congress, but as SDI will not close Pandora's box of bad nuclear genies, Congress will not do away with SDI. To restrict France to a purely negative attitude is not realistic.

More importantly, SDI is a decisive, technological and industrial process. It is an instrument of industrial policy—one that will channel important, massive credits. When one confronts that aspect of SDI, one should distinguish four elements, some of them contradictory, in the French attitude. On the negative side, there is the fear that SDI research will further reduce, by its spin-off, the level of competitiveness of European industries. Europe is lagging behind. Because of SDI, it will lag behind still further.

There is also the fear that none of the myriad hopes entertained by some Europeans, particularly in the UK, will materialize; that basically the protectionism of US firms, the distrust of the US Congress and of the Pentagon, will limit the dissemination of information and will leave the Europeans very much disillusioned by false hopes.

But, at the same time, and this is the positive aspect of the issue: there is the feeling that space industry, and high technology in particular, is very competitive; that the train should not leave without us; that, in fact, on the key aspect of 'leakage', we might be better than some of our European counterparts—to put it bluntly, that we might be better (at keeping secrets) than the Germans are, given the extensive contact between the two Germanies.

In fact, that mixture of temptation and fear has taken on a name of Greek origins: Eureka. The goal of Eureka, which is not presented as an anti-SDI project or as anything that should compete with SDI because it is specifically civilian and not military, is to transcend the technological Balkanization of Europe, to present a common European answer to the technological challenges which SDI accelerated.

Undoubtedly SDI jolted Europe—making it brutally aware of the hi-tech problem. Secretary of Defense Weinberger's letter of 26 March 1985 asking for an allied response within 60 days to a US invitation to co-operate in SDI research added to this shock. SDI was then considered to be a technological challenge which had to be met one way or another. The seeds of Eureka were sewn by President Mitterrand in a speech in The Hague in 1984.[4]

Eureka is aimed at assuring European technological independence in an area which promises to be vital in the future. In no way is it a European SDI; since the military will benefit from the spin-offs of what is primarily a civilian project, Eureka is rather the inverse of the US SDI. As stated by Foreign Minister Roland Dumas in the National Assembly, 'The Eureka project is primarily civilian in spirit. Results of advanced research, however, in fields such as micro-informatics, optical technology, robot technology or high-energy physics, will also contribute towards increasing Europe's responsibility for its own security'.[5] There is thus a common technological 'trunk' which serves both the civilian and military sectors. The intention is that this scientific and technological base must be supported and also enlarged and enriched.

Although the seeds for this idea of pooling European technological knowledge were planted earlier, the process itself was initiated by the letter of 15 April 1985 to France's European partners from Foreign Minister Roland Dumas, seeking their reaction to such a programme. France's initial goal for Eureka was first to define at the ministerial level the future of European research, but also to stimulate European R&D, and finally to define programmes of action. All France's partners responded favourably, and reiterated their support for Eureka during the June 1985 European Community Milan summit.

In fact, it is not surprising that under the Socialist government there was a mixture of rejection and temptation. Under President Mitterrand, the French foreign and defence policy has been to a large extent very Gaullist. At the same time, it was under socialism that the discovery of the centrality of private industry was made. In his memoirs, when he speaks of challenges, President Mitterrand invokes an industrial Waterloo or an industrial Austerlitz. International competition is today usually synonymous with industrial competition, an essential element of the SDI debate.

III. Conclusion

To conclude, one should not underestimate, first, the short-term diplomatic consequences of SDI on European co-operation. It has, unfortunately, served as an accelerator of Franco-German divisions. It would be unfair to put the blame on the US alone. Responsibility for the present blocking of Franco-German co-operation must be shared among the French, the West Germans and the Americans. But clearly, the presentation by the US of SDI was not helpful. It was, in fact, contrary to the proclaimed support of progress in the European defence of Europe. There was a contradiction between what the US had started to promote in 1982, and the real diplomatic impact of SDI on Franco-German relations.

More profoundly, the SDI debate in France is a catalyst of France's own anxieties. Coming after the Euromissile quarrel, SDI shows that the rather comfortable world on which France had based its security for more than 20 years is slowly fading away. During the 1960s, protected by US nuclear supremacy, by the Federal Republic of Germany's absolute stability, we could promote an independent way within the alliance without endangering our security.

Today, between SDI and the emerging technologies of conventional weapons, between Washington's dreams and the calculus in Geneva, the two values of independence and security may not be pursued together. More security tomorrow may mean less independence, and that is the key to the SDI debate and to the way the French are perceiving it. It puts into question the traditional world on which France has based its security. The ultimate temptation for France would be to distance itself from the strategic implications of SDI and its emphasis on defensive deterrence, whether buying fully its industrial implications or not. Such a position cannot be easily sustained, because ultimately it is contradictory.

But above all, what is at stake ultimately for France is its self image and the close linkage created in this country between independence, security and absolute reliance on nuclear deterrence. For France, a strategy of deterrence based on the nuclear component has been in many respects ideal. The twinning of deterrence and détente brought to the Vth Republic a protective space within which France could pretend to possess cards that were at least equal to those of any other power. The nuclear weapon, factor of uncertainty for all, constituted for France an equalizing factor, which allowed it to hide material inferiority *vis-à-vis* the superpowers and to widen its margin of diplomatic manoeuvre within the Western alliance. Today, the attack on the sanctity of nuclear deterrence does not come solely from a Soviet Union ever willing to exploit the growing frustration of Western public opinion *vis-à-vis* the nuclear arsenal and to benefit from the stark nudity of a conventional imbalance that greatly favours her. The Reagan Administration has also joined the club of the 'revisionists' trying to define a world beyond nuclear deterrence.

In reiterating her belief in a safer world, thanks to nuclear deterrence, France is proven right by history. Wherever nuclear deterrence has existed war has been prevented. Men and not weapons make wars. Violence in the world comes not from the existence of nuclear weapons but from traditional arms conflicts, from ideological and political factors and from the continuation of dramatic international economic disequilibrium. Is France fighting a rearguard battle against irresistible, diplomatic, psychological and technological evolutions? What seems destabilizing today for the French is the impending prospect of a superpower agreement that contains the growing threat of French disarmament by obsolescence and through the denuclearization of Europe. Despite the present position of the two superpowers on the non-inclusion of medium-sized nuclear powers in the negotiations, who can say with absolute certainty that at some point the US, USSR and the European non-nuclear powers may not be tempted to strike a global deal that will necessarily include French forces? For the French, the ideal dialogue between Washington and Moscow has to be slow and difficult without spectacular crises or dynamic breakthroughs. So far pressures on the French independent nuclear force are external and have not impinged on French public opinion. But at a time when the necessary modernization of French nuclear deterrence will require economic sacrifices such international questioning of the moral, strategic, social and technological bases of the French nuclear choice can only be disturbing.

Ultimately what is at stake is the capacity for a medium-sized power to keep an independent policy while preserving its security. Now that under the impulse of President Reagan the United States has joined the club of the 'revisionist' powers, France's efforts at preserving the *status quo* will prove increasingly difficult.

Notes and references

[1] Dr Moïsi is Associate Director, IFRI; and Editor, *Politique étrangère.*

[2] Mitterrand, F., *Reflections sur la politique exterieure de la France* (Fayard: Paris, 1986)

[3] Since Chernobyl, however, and a succession of minor incidents at various French nuclear plants, dissent over nuclear energy policy has grown in France.

[4] Speech by Mitterrand in the Second Chamber of the States-General, The Hague, 7 Feb. 1984.

[5] 'Dumas: Eureka may have military repercussions—Gorbachev in Paris?', *Atlantic News*, no. 1728 (14 June 1985), pp. 3-4.

Paper 10. Tactical missile defence

Hubert M. Feigl[1]
Stiftung Wissenschaft und Politik (Institute for Science and Politics), D-8026 Ebenhausen, Federal Republic of Germany

I. Introduction

For many years the contents of the 1972 Anti-Ballistic Missile (ABM) agreement have been subject to an insidious erosion which visibly jeopardizes its continued existence. Soviet activities which may be related to the development and possibly nation-wide deployment of new missile defence systems have already lead to US apprehension that the Soviet Union is about to break out of the agreement.[2] On the other hand, the USA with its Strategic Defense Initiative (SDI) has committed itself to a course that, assuming positive progress of the SDI research programme, will sooner or later make abrogation of the ABM agreement inevitable. Of course, some testing activities could contravene the limitations of the agreement long before a decision is taken about the introduction of SDI systems.

The process of erosion is characterized not so much by sensational violations of the agreement that justify apportioning of blame to one side or the other. Rather, erosion of the Treaty is marked by a multitude of US and Soviet development activities not covered by the agreement, or only insufficiently dealt with, although they are relevant to its maintenance. Technically 'adjacent' areas of weapon development offer an obvious opportunity for the further development, testing and introduction of ballistic missile defence (BMD) components relevant to the agreement. Among the weapon categories which are interesting in this context are anti-satellite weapons and particularly the tactical missile defence weapons dealt with in this article. The latter are intended for defence against non-strategic missiles and are as such beyond the 'reach' of the ABM agreement. Accordingly the US was rather quick to recognize the advantages of a 'dispersed' development and possibly also earlier regional deployment of anti-tactical ballistic missile (ATBM) systems.[3]

The ABM agreement does not prohibit the transfer of missile defence systems and their components by the signatory states, as long as these are not strategic BMD systems or their components. Provided these conditions are fulfilled, development of tactical missile defence systems from air defence systems is also allowed.

Drawing the dividing line between an ATBM and an ABM capability is difficult in practice because of the fluidity of the boundary in cases where technical criteria rather than criteria of intended use are applied. It is therefore impossible to avoid a certain margin of interpretation regarding functional and operational application of ATBM systems, which in principle also include air defence weapon systems. For the time being, ground-to-air missiles and radar components would have to be used as defence against tactical ballistic missiles. These are also necessary to combat aircraft and missiles with aerodynamic flight profiles. The important factor in this context is that the limited ATBM capability of these systems may be accompanied by a marginal ABM capability. However, their military significance is, at least, doubtful. On the other hand, it is impossible to deny the ABM capability and the military relevance of the basically similar high performance missiles and radar components which are indispensable for the terminal phase of a strategic missile defence.

It is thus plain that these delineation problems create 'grey areas' in weapon development, which could be used to 'erode' the ABM agreement. Conversely there is no concrete possibility of interpreting the ABM agreement in such a way that it could be extended to ATBM or even air defence systems. In order that this situation may be better understood, the following will deal with the technical preconditions, the current state of development and the future perspective of tactical missile defence in Western Europe and the Soviet Union.

II. Tactical missile defence for Western Europe

It was obvious as early as 1944 just how problematic defence against ballistic missiles really is. Even at that early stage it was realized that it was impossible to protect the target areas of the German V-2 rockets, which today would be known as short-range ballistic missiles. The only reason why the impact of these 'wonder weapons' was limited was that their accuracy was poor and the warhead detonators often failed. In post-war years when the leading nuclear weapon powers (USA and USSR) set about equipping accurate long-range missiles with nuclear warheads, creating weapons with an enormous destructive potential, the demand for suitable means of defence became all the more urgent. The development of missile defence weapons was tackled only by the two superpowers, who felt obliged to protect their territories against nuclear missile attacks. Building on the first concept studies shortly after World War II, both sides developed ABM systems during the 1960s, which were based on the principle of guided air defence missiles and which could only engage ballistic targets with nuclear warheads. Although, in those days, the United States obviously had a considerable lead in development over the Soviet Union (which had at a very early stage decided on ABM deployment), the US Government could not make up its mind during the 1970s to introduce the inflexible and also very vulnerable Safeguard ABM system on a nation-wide basis.

At the end of the 1960s the European NATO partners, which from the very beginning had foregone any missile defence developments of their own, declared themselves to be categorically against an ABM system.

Considering the peculiarities of their security situation they could see even less reason to adopt inadequate Safeguard ABM technology. When at last the two superpowers signed the ABM agreement in 1972, influenced, not least, by the low cost-effectiveness of their missile defence technology, the subject of European missile defence disappeared into the background for many years.

It was really only President Reagan's famous announcement on 23 March 1983, resulting in a new political initiative, SDI, which set the Europeans thinking again about missile defence. It instilled fear in the Europeans that only the USA would be protected against a missile attack. This fear was not unfounded, as the aim of the project was unmistakably to neutralize the main threat to the USA: long-range nuclear missiles. The European NATO countries were not slow to comprehend that the new possibilities of non-nuclear intercept of ballistic missiles in a layered missile defence system, as conceived in SDI, could also in principle have attendant advantages in the context of regional missile defence.[4] There are, however, weighty political, military and not least budgetary reasons against the European off-shoot of SDI which was being demanded by the supporters of the European Defence Initiative (EDI). There is certainly a much better chance of a European missile defence based on an interdependent system coming into being, as has indeed been promised by the USA. This envisages combining a space-based missile defence component—which can be introduced globally and which will be developed within the framework of SDI—with a regionally deployable missile defence component, for use in Europe.[5] Of course, this means that it is also burdened with all the imponderables of the US development programme and its implications for NATO strategy and arms control policies.[6]

Seen as a whole there are unlikely in the foreseeable future to be any easy solutions to the problem of neutralizing the nuclear options of the Soviet Union against Europe, even if the outcome of the SDI programme were successful. The threat to Europe is too multifarious and too difficult to quantify for those politically responsible to assess it conclusively at this juncture.

The threat to Western Europe

Western Europe is threatened today by a comprehensive arsenal of ballistic missiles covering all ranges from ICBMs (e.g., SS-11s and SS-19s) and various SLBM versions to the various types of intermediate- and short-range missiles. The 441 mobile SS-20 Saber intermediate-range ballistic missiles (IRBMs) with a range of up to 5000 km form the greatest regional threat. More than half of these triple warhead units are said to be aimed at targets in Western Europe. On top of this there are more than 100 older SS-4 Sandal single-warhead missiles (range 2000 km) which have not yet been replaced by SS-20 missiles.

Following the most recent Gorbachev initiative, the prospects that at least the SS-20 arsenal (apart from some 100 warheads remaining in the Asian part of the Soviet Union) can be eliminated through negotiation have improved considerably. It should, however, be borne in mind that the Soviet ground force within the Warsaw Pact still includes at least 1300 other nuclear short- and intermediate-range missiles with ranges of between 120 and 900 km, which can partially take over the tasks of the SS-20s.[7] This missile potential is composed of the old Frog-7 (range 70 km), the SS-1c-Scud B (range 300 km) and the SS-12 Scaleboard (range 900 km), which for a while have been in the process of replacement by the modern SS-21 Scarab (range 120 km), the SS-23 Spider (range 500 km) and the SS-22 Scaleboard (range 900 km).

It is assumed that the Soviet Union's medium- and short-range missiles alone are in a position to cover all important military targets in the NATO central area of command. As is generally known, disarmament talks regarding these missile categories are envisaged although at the time of writing it was completely open as to whether drastic reductions would result. It is also unclear in what form the broadly based modernization and extension programmes will be continued. Seen from today's point of view it could be that the already perceptible focus on development of long-range weapons with an aerodynamic flight profile may become even more pronounced. The Soviet Union is currently not only modernizing its total missile potential in the intermediate and short-range categories, but also developing improved cruise missiles.[8] On top of this there are the new stand-off weapons, which can be used against targets in Western Europe from distant aircraft.

Many Western defence analysts see a further danger. Due to the increase in accuracy of long-range missile systems the Soviet Union could for the first time use conventional and chemical warheads against distant targets with military effectiveness. Of most prominent significance in this context would be the ballistic short- and intermediate-range missile forces, especially the new SS-21, SS-22 and SS-23, which could be equipped with modern warhead technology (e.g., fuel air explosion, but also smart sub-munitions). The reason is that existing NATO air defence would be ineffective against these missiles.[9] It is feared that this could very quickly result in qualitatively new conventional engagement options for the Soviet Union. The priority targets of such presumed attacks would be nuclear weapon systems, special weapon depots, air defence systems, especially radars, airfields, large concentrations of troops, operational reserves as well as command and telecommunication centres.

There is even a fear within NATO that without appropriate countermeasures on the part of the West the Warsaw Pact would in this way be able to achieve a conventional first-strike capability that could impede defence operations decisively and furthermore undercut the timeliness of nuclear escalation.[10]

Extended air defence

There is therefore a general conviction that the time has come to respond, particularly to new challenges in the non-nuclear segment of the threat spectrum.[11] Several study groups at NATO level as well as on a national level are already examining suitable measures. Besides passive measures such as the hardening and camouflaging of targets, the most interesting countermeasures are those which could be achieved under the conceptual framework of an 'extended air defence system'. What is called for is an exclusively non-nuclear, multi-capable defence, which includes the use of ATBMs.[12] The ATBM potential should above all neutralize the threat from non-nuclear short- and medium-range missiles. The problems of an active defence against nuclear ballistic missiles covering missiles of all range categories should be reserved for a later solution within the SDI framework. Thus the concept of extended air defence can be realized independently of the future progress of the SDI programme. Conversely, should SDI be realized, the option of an interdependent expansion of the defensive umbrella would remain intact.

The development of a European ATM/ATBM system is not hindered by the ABM agreement and should gradually evolve from the existing structure of NATO air defence. The aim of this development is not, however, the realization of a comprehensive area defence, but primarily the defence of high-priority single targets.

The main task of 'extended air defence' would be the neutralization of the conventional threat based on air-breathing and ballistic weapon systems. But in the absence of suitable means for identifying conventional and chemical warheads, defensive actions must be extended to take nuclear weapons into account. At the present time the Soviet Union can resort to ballistic missiles with a range of up to 1000 km and a multitude of air-breathing systems in order to safeguard her non-nuclear engagement options.

The ATBM component of an extended air defence system should, of course, be capable of intercepting any non-nuclear ballistic missiles. Its effectiveness will primarily be measured by its success in combating the modern tactical missiles SS-21, SS-22 and SS-23. As the ballistic trajectory of these single-warhead missiles is largely (SS-22 and SS-23), or entirely (SS-21), within the atmosphere, there is no danger of the aggressor using decoys and other means to saturate the missile defence. The defender has, however, the disadvantage that missile defence weapons which develop their full potential outside the atmosphere are of little use. That is why an ATBM defence must rely mainly on other types of defensive weapons; generally ground-to-air missiles. Consequently, as target engagement must be carried out from those areas which are to be protected, a boost-phase defence can scarcely be contemplated, only the possibilities of a 'late' mid-course defence and above all a terminal-phase defence can be exploited. A relatively late response further impedes the defensive task. Furthermore, as multiple engagement of ballistic targets is limited to the shoot-look-shoot principle, it is more difficult to achieve a highly efficient missile defence.

The approach speeds of warheads and missiles of the medium- and short-range categories are relatively low and thus compare favourably with the interception speed of ground-to-air missiles. This is of considerable importance as warheads and missiles should be intercepted as early as possible, that is, as far as possible from their targets. The interception distances determine the size of the area defended by an ATBM battery, and thereby also the opportunities for a more rational usage of several ATBM positions in order to increase the area covered or increase the level of protection of specific targets.

The degree of defence effectiveness may also be called into question regarding the relatively short total flight times of the short-range missiles (e.g., 3 to 4 minutes for a SS-21 with a range of 120 km). That is why, when defending against intermediate- and short-range missiles, earliest possible target detection and target acquisition are at least as important as the capacity for acceleration and the final speed of the interceptor missiles. Ultimately the prospects for successful missile defence depend decisively on whether the time available for target engagement can be used optimally. The basic precondition for a good defensive capability is a high-performance early-warning sensor system. With early-warning information gained from, for example, special aircraft, highflyer drones or other unmanned airborne platforms, the target detection ranges of ground-based radar installations may, for instance, be tripled and, with correspondingly powerful weapons, the interception range more than doubled. This is principally a case of prompt handling of very small target signatures (e.g., small radar cross-sections of warheads). Sensor systems for surveillance, target tracking, acquisition and kill assessment can considerably facilitate defence.

However, all these efforts are in vain if there is no highly reactive command and control system, capable moreover of functioning largely autonomously. This is especially true if target intercept is also to be extended to ballistic missiles launched from only a short distance. The use of non-nuclear defences is, as a general principle, important in this context.

Of course, the destruction of a ballistic target through explosion or collision ('weapon kill') demands an extremely precise intercept, placing the highest demands on the guidance system. It may be sufficient to deflect an incoming warhead from its target course through a near-by explosion within the atmosphere ('mission kill'). This defensive procedure will, however, be more promising when non-nuclear warheads are to be intercepted since they depend much more on precise targeting. Intercept of nuclear warheads would *a priori* require 'weapon kill' and moreover at the greatest possible altitude in order to minimize the risk of collateral damage should a nuclear explosion be triggered (e.g., by salvage fusing).

Favourable conditions for the destruction of a ballistic target by direct hit or collision occur when the ground-to-air missile is used as a so-called 'head-on interceptor', for instance with the help of a beam rider technique, that is, when the defensive missile can be fired along the predetermined flight path of the attacking ballistic target. This defensive procedure does, however, demand that the ATBM position is in the immediate vicinity of the target being defended. The kill rate of missile defence will drop off considerably when more distant targets are to be protected. For this reason it is difficult to achieve more than a point defence capability with early generations of ground-to-air missiles.

At short notice a missile component that provides at least some contribution to an 'extended air defence' can only be achieved by a modification of already available anti-aircraft systems. Of primary relevance in this context is the Patriot mobile air defence system, which is being deployed with the US army in Europe as well as with the German and Dutch air forces. By the mid-1990s there should be a total of 82 Patriot units with a total of more than 6000 missiles available in Western Europe. The Patriot missile reaches supersonic speeds (1.6 to 2.0 km/s.) and is equipped with a special warhead for conventional target engagement. Its maximum range of interception is said to be more than 70 km against airborne targets. The decisive advantage of the Patriot lies in the fact that multiple target engagement can be guaranteed with a single phased-array radar. Every firing unit can track up to 50 targets and launch up to five missiles simultaneously.

The Patriot was originally conceived and developed as an ATBM-capable air defence system under the system name SAM-D. However, this ambition was later considerably moderated, since the air defence capability of the Patriot was regarded as satisfactory. This means that the SAM systems currently stationed in Europe are only capable of intercepting aircraft and cruise missiles with relatively large radar cross-sections. In order that the Patriot can be used within the framework of an extended air defence the multi-purpose radar and associated software must be adapted to ATBM requirements.

A two-stage modification of the Patriot system has been suggested within the framework of the so-called PAC programme (PAC = Patriot ATBM Capability).[13] The aim of PAC I is primarily to fulfil the basic requirements for a successful mission kill of SS-21, SS-23 and similar missiles. This package of measures includes new software and changes in radar technology which could be achieved within the next few years. As early as September last year the USA carried out its first successful test of an improved Patriot against a tactical ballistic missile of the Lance type.[14] PAC II, which could be completed by the beginning of the 1990s, is ultimately supposed to give the Patriot system warhead kill capability. Besides further software improvements, significant hardware alterations are required to this end. Among these are more effective warheads, new fusing systems and probably also an additional booster stage to increase the speed and range of the missile.

By means of the PAC programme the Patriot batteries could relatively quickly be provided with a self defence and point defence capacity for ATBM tasks. It must, however, be assumed that each firing unit cannot defend more than 200 km². There are thus clear cost-effectiveness limits to any attempt to achieve a comprehensive area defence by use of additional defence batteries. In order to create better operational conditions the experts are calling for the development of a completely new ground-based missile defence system instead of modification of the Patriot system.

The question of which anti-aircraft system the Patriot is supposed to complement for engagement at medium altitudes has not yet been answered within NATO. Moreover, it is particularly dubious whether the Hawk air defence system can still be considered for an extended air defence system. Hawk was introduced as early as 1963 with a number of subsequent modifications (maximum range of interception of airborne targets 30 km).

At best the Hawk could be equipped with a limited defence capability against cruise missiles. In an ATBM role, however, they would have great difficulty even with respect to their own self defence.

In the meantime, extending ATBM capability from existing air defence systems is only regarded as an intermediate step. In some European countries and not least in the USA the conviction has grown that further steps in development must be taken in the course of the replacement, planned for the 1990s, of the older generation of anti-aircraft missiles. In general, only those systems which can counter all types of missiles should be considered as a replacement. The development of a new medium-range SAM system with limited ATBM capability is of particular urgency in this context.[15]

In France, for instance, one consortium of companies is busy with the system SAMP-90 (Sol-Air Moyenne Portée) and its naval version SAAM-90 (Sol-Air Anti-Missile). Defence against aircraft, cruise missiles and stand-off missiles is what is aimed for in the first place. For the time being, the tactical demands of France do not include any ATBM capability. The SAMP-90 could, however, be equipped with such a potential with low development risks and at a small cost. Preliminary investigations have shown that even the standard version would be capable of repulsing the Soviet SS-21 short-range missiles. The most important SAMP-90 system components are the Aster-30 missile defence missiles (range 30 km) and the phased-array multi-purpose radar Arabel, which is capable of tracking 50 targets simultaneously and guiding 10 missiles to their targets.

A group of companies in the Federal Republic of Germany is working on the MFS-2000-System (Mittleres Fla-Rak System or intermediate ground-to-air system), which has been renamed TLVS (Taktisches Luftverteidigungssystem or tactical air defence system). The West German medium-range SAM project is still in a relatively early stage. In accordance with tactical requirements, comprehensive defence capabilities will probably be demanded which can be put into effect against fast low-flying aircraft, missiles and helicopters. The main task of the TLVS is air defence against low- and medium-altitude saturation attacks as a complement to the Patriot. The realization of an extended air defence does, however, require at least a limited ATBM capability.

The TLVS concept envisages a highly mobile SAM system on wheeled terrain vehicles and equipped with a very high degree of fire-power. Faced with varying operational tasks, the system must be able to use two types of missile with differing system configurations. With this in mind a combination has been suggested comprising one missile with a range of 30 km, which exceeds the speed of sound and is equipped with an active radar search head, and a smaller, extremely versatile missile with simple sensor equipment, which compensates for a maximum range of only 10 km by reaching much higher hypersonic speeds. The two types of missile together should have a limited ATBM capability against conventionally armed ballistic missiles, involving a self-protection capability and a development potential for target protection.

In the medium term the TLVS programme can also be complemented by other development programmes. A considerable improvement in the cost-effectiveness of a tactical missile defence is expected with, for instance, anti-aircraft guns using newly developed manoeuvrable shells against conventional warheads. The possibility of countering the re-entry vehicles of ballistic missiles from fighter aircraft with hit-to-kill guided weapons was also considered for a time.

Two comparable air defence systems are being developed in Europe at the moment: the French SAMP-90 and the German TLVS. Individually or combined they could form the basis of a future medium-range SAM system for NATO. Further contributions are expected from other NATO countries, particularly Great Britain and the USA, which have been working on similar projects for a long time.

The form and extent of a US participation in extended air defence has not been made clear as yet. Particularly in the case of ATBM applications the USA could introduce a range of relatively advanced developments, the majority of which are components of the SDI programme. Thus the Exoatmospheric Reentry Vehicle Interception System (ERIS) comes into consideration for ATBM mid-course defence. This system destroys attacking warheads and missiles by collision (maximum altitude of interception more than 120 km, maximum slant range more than 300 km). The High Endoatmospheric Defense Interceptor (HEDI) programme could be employed in the upper layers of the atmosphere. In this case the HEDI defence missile would be armed with a conventional explosive charge (maximum altitude of interception around 50 km, maximum slant range around 60 km). The USA developed the Low Endoatmospheric Interceptor System (LEDI) for the bottom layer of a terminal phase defence. This is designed to intercept conventional warheads at a distance of about 6 km. Furthermore, the Flexible Lightweight Agile Guided Experiment project (FLAGE) that builds on the Small Radar Homing Intercept Technology programme (SR-HIT), must also be mentioned. The latter two projects contribute to the development of new guidance systems for hit-to-kill weapons. For the time being, however, it remains to be seen whether these developments, which should see fruition at the beginning of the 1990s, can be applied within a European missile defence system.

Other SDI combat techniques envisaging the application of more exotic technology and mainly for use outside the atmosphere can probably not be matured this century. Furthermore, only a few of them can be considered for tactical missile defence within the conceptual framework of extended air defence. Only the ground-based, and with certain modifications even the air-based high energy laser (HEL) have relatively good operational possibilities. Space-based laser weapons and ground-based lasers with mirrors in space are, on the other hand, typical components of the SDI concept, mainly suitable for combating targets in space. HEL missile defences that do not rely on space components are, of course, most useful when a large target proliferation has to be dealt with in a very small space of time. Ground-based ATBM laser weapons with sufficient power should be able to intercept not only approaching warheads but also missiles at a distance of 100 to 200 km. However, the range of these beam weapons may be limited due to unavoidable energy losses in the atmosphere.

It is also questionable whether ground- or air-based 'electromagnetic guns', which are a component of the SDI programme, can be used. These weapons are said to use either guided or unguided small, high-speed projectiles at a high firing rate. However, intercept ranges sufficient for ATBM purposes can only be achieved with guided projectiles. Because of the extremely high technological standards which have to be met, the perfection of such guided projectiles is particularly questionable. That is why any tactical missile defence will have to rely mainly on ground-to-air missiles until long after the year 2000, even if the SDI programme progresses uninterrupted.

III. Soviet developments in tactical missile defence

The concept of defence has always had a prominent position in Soviet military thinking. According to Soviet doctrine the outcome of war is determined not only by military victory on the battlefield but also by the ability to preserve one's own territory against destruction.

At a very early stage an independent military division for national air defence was built up; and before long the task of strategic missile defence was placed there. Of major interest in this area were defence systems suitable for protecting Soviet territory or at least important parts of it from 'strategic' missile attacks. The development of weapons to counter 'tactical' missile attacks, with the capability of defending Soviet troops on the territory of Warsaw Pact countries by means of 'advance stationing', was evidently postponed for a long time. Logically, therefore, Soviet ABM development concentrated on combating long-range ballistic missiles capable of reaching Soviet territory. Among these are the US intercontinental and submarine-launched ballistic missiles (ICBMs and SLBMs), as well as the various strategic missiles stationed in Western Europe by NATO or the US, Great Britain and France and in Eastern Asia by the People's Republic of China. The 'European component' essentially consists of British SLBMs of the type Polaris A-3 (range 4600 km), French SLBMs of the types MSBS M-20 (range 3000 km) and MSBS M-4 (range 4400 km), French intermediate-range ballistic missiles (IRBMs) of the type SSBS-S3 (range 3500 km) and US/NATO medium-range missiles of the type Pershing II (range 1800 km).

On top of these come the so-called tactical missiles such as NATO's Pershing 1A (range 160-720 m) and Lance missile (range 110 km) as well as the French Pluton (range 120 km), which although incapable of reaching Soviet territory can reach the territories of allied Eastern bloc countries.

In order to achieve comprehensive protection of the Soviet Union it is important that any missile defence system which is effective against attacking intercontinental strategic missiles is also suitable for combating longer-range continental strategic missiles. Moreover, some components of a ground-based terminal-phase and mid-course defence, central components of any layered missile defence, also have in principle an ATBM capability which can be used to defend against intermediate- and short-range missiles. The latter field of activity can, however, also be covered by ATBM systems specially developed for this purpose. These systems are more suitable for forward deployment as well as being capable of integration in the 'overall system' that is also effective against intercontinental strategic systems. It is not useful to judge the Soviet ATBM development by the relatively restricted NATO standards which are applicable today. This is because the Soviet Union is active in both areas of development, relevant technology is easily interchangeable and progress in development is facilitated by a more unified space defence programme. On top of all this comes the ever more difficult problem of finding a cut-off point between 'anti-tactical' and 'anti-strategic' BMD capability and the fact that one cannot, in principle, exclude the possibility that even advanced air defence systems suitable for conventional defence may have limited ABM characteristics.

An assessment of Soviet willingness to keep to the ABM agreement is necessarily subject to some uncertainty. This will be reduced when further system features can be recognized at a relatively advanced stage of development.

Right from the start the development of anti-missile systems in the Soviet Union was obviously beset with great difficulties and was characterized more by persistence and determination than by quick results which could be put to practical use.[16] The first programmes were probably initiated in the 1950s. In those days even the Soviet Union had no option other than to embark upon the unattractive road to nuclear missile defence by means of guided ground-to-air missiles. In the early 1960s technically very inadequate defence systems were deployed outside Leningrad and Tallinn, but were soon to be dismantled again. Even the ABM-1 system which was built up around Moscow between 1968 and 1969 was still based on a relatively low level of missile defence technology. This was a single defence system served only by the Galosh interceptor missile. The Moscow system, because of its inadequate radars, could have been overcome with relative ease by strategic missiles which would have brought multiple warheads, decoys and other means of penetration into operation. Thus the Galosh system was at best suitable for defending against limited missile attacks which, even then, could have been carried out by the smaller nuclear powers.

Improved Moscow defence

The ABM system for the greater Moscow area, as is well known, remains the only operational BMD system in the world. The long-awaited extension into a two-layered missile defence evidently got under way in 1978. Since 1983, the improved silo-based long-distance ABM SH-04 Galosh (range over 300 km) for exo-atmospheric operation has been deployed.

This interceptor is complemented by the SH-08 Gazelle short-range ABM (range over 80 km), which is also silo-based. This is a high-velocity missile for endo-atmospheric operation. Further components are the Pawn Shop missile guidance radar and the Flat Twin target-tracking radars, which are phased-array radars and transportable. In addition, the gigantic Pushkino radar, probably for ABM fire control, is also available. This radar complements or replaces the older Dog House and Cat House battle-management radar systems. For many years the buildup of a multi-layer early-warning system has been pushed along in parallel to this with great effort. It consists of satellites, radars with over-the-horizon capability, and long-distance radars of the older Hen House type as well as the newer Pechora type.

The new ABM-3 system is also designed for using only interceptor missiles with nuclear warheads. This means that its operational capability is scarcely likely to exceed that of the older US 'Safeguard' system which was scrapped shortly after the ABM Treaty was signed. This is certainly the case if ABM-3 is only to be equipped with the 100 ABM launchers allowed by the ABM agreement and the reloading capability dispensed with. For the time being there are no definite indications that the Soviets will exceed the permitted number of launchers. There are, however, suspicions that the SH-08 silos may be reloadable in contravention of the agreement.

A nation-wide defence?

The opinion within the Pentagon is that the SH-08 and SH-04 defensive missiles, together with the Flat Twin tracking radar and the Pawn Shop missile-guidance radar, also form an autonomous, transportable defence system, the components of which could be produced in large quantities, enabling the rapid development of a missile defence with area coverage.[17] The advanced state of assembly of urgently needed radar networks for target detection and tracking is believed to be a further indication of such an intention.

If this means that the architecture of a nation-wide operational missile defence system is in preparation it would, in principle, also include an ATBM capability. Judging by its performance characteristics, an extended stationing of ABM-3 would be considered more for the protection of hardened targets such as missile silos or underground command and control centres. Its use as an ATBM system for the protection of soft targets such as concentrations of troops is likely to be of secondary significance by comparison. The fact cannot, however, be ignored that the ABM-3, in its capacity as an autonomous defence system, would form an important intermediate step towards a non-nuclear missile defence which would offer better operational options right from the start.

It is very unclear how the results of Soviet developments in the field of exotic anti-missile technology could be applied to an ATBM system. It is most likely that the extensive research programme in the field of high-energy lasers will find expression in operational laser air defence systems with an ATBM capability. Otherwise it can be assumed that technical and operational constraints also apply in principle on the Soviet side and will lead to programmes similar to those in the West. Ultimately much will depend on whether the various areas of application of exotic weapon developments can in their entirety be included in arms control agreements.

For many years NATO has been watching the development of Soviet air defence missile systems with great interest. These are also ascribed a non-nuclear ATBM capability. It should be noted here that the Soviet Union has, possibly for much longer than NATO, been planning to assemble components of a tactical missile defence within the framework of an integrated air defence.

Among the various SAM systems of the national air defence forces only the SA-5 and primarily the newer SA-10 are ascribed limited missile defence capability.[18] A more specific ATBM capability can only be expected from the newly developed SA-12 air defence system, which is evidently intended for action within the Soviet ground forces.

The SA-5 Gammon strategic long-range air defence system was ready for operation as early as 1967 and has since then been subject to several modifications. With a maximum intercept altitude of about 30 km and a maximum slant range of 300 km, it is the most important strategic air defence system the Soviets have today. In past years at least 2050 launchers for SA-5 missiles have been introduced. This air defence system can use nuclear and conventional warheads. The Gammon missile is supposed to have been tested in an ATBM role some 50 times as early as the 1970s. The ATBM capability of this relatively old defence system is regarded as rather marginal today. The possibility cannot, however, be excluded that there will be a further modification of the SA-5 system. In conjunction with other ATM-capable, short-range SAM systems this could considerably increase the effectiveness of the air defence.

A new generation of intermediate-range air defence missiles has become known under the name of SA-10 Grumble. After a long period of development, deployment of this SAM system in the Soviet Union started in 1980. By the end of 1986 more than 70 sites were ready for operation and at least 20 sites were under construction. It is assumed that by now the Soviet Union has deployed considerably more than 735 launchers for the stationary version of SA-10A. On top of this the SAM system is sufficiently transportable to permit relocation. As these defence missiles are obviously intended for vertical take-off they can also be launched from fixed silos.

It has three transportable 'advanced' radar systems at its disposal with which it is believed to be able to find and track even low-flying targets such as cruise missiles, which have very small radar cross-sections. According to US sources a mobile version of the SA-10 is also being introduced, capable of vertical launch from special wheeled vehicles. The missiles are held in readiness in groups of four on the launching and transport vehicles. The phased-array missile-guidance radar is also said to be accommodated on one of these vehicles. The missile itself is characterized by a high-acceleration capacity, a high terminal speed of Mach 6, variable engagement height (300 m to at least 50 km) and a long range (around 100 km). The system can launch conventional and possibly also nuclear warheads. The main task of the SA-10 is undoubtedly the combat of manned and unmanned aircraft at all altitudes. According to US sources these weapon systems may also be intended for defence against long-range cruise missiles. The SA-10s are also considered effective against comparatively slow intermediate and short-range ballistic missiles. The same sources do not exclude a limited defence capability against some types of long-range ballistic missiles, primarily SLBMs. [19] The strategic BMD capability of the SA-10, however marginal, is felt to seriously call into question Soviet compliance with the ABM agreement.

In the opinion of the Pentagon, the new SA-12 air defence system represents the most important step yet on the part of the Soviet Union towards a general ATBM capability.[20] It is noteworthy that this aim is being pursued, as it is by NATO, within the framework of an integrated air and missile defence.

The SAM system is expected to be able, with good prospects of success, to combat not only aircraft at any altitude but also cruise missiles and short-range ballistic missiles. A high-speed missile with an altitude range between 100 m and 30 km (maximum operational range around 100 km) has been deployed. In addition to this, the radar-guided missile can, besides a normal warhead, apparently also be armed with a special warhead for engagement against ballistic targets. The system's missile-launching vehicle, moving on tracks, carries two cylindrical container launchers and a telescope-like missile-guidance radar. The weapon system also includes escort vehicles for the target-tracking and fire-control radars as well as vehicles for the transport of further missiles in twin containers. Because of different task positions the new mobile SA-12 system exists in two versions: the SA-12 Gladiator and the SA-12 Giant.[21] The Gladiator which is already being introduced is, in the opinion of Western observers, intended primarily to combat airborne targets, although in principle it should also be suitable for missile defence. It is assumed that it will replace the SA-4 Ganef as the standard SAM system on the army and front levels of the armed forces of the Warsaw Pact. The incompletely developed SA-12

Giant version, which is also mobile, can probably take over more extensive ATBM tasks.

The SA-12 has already been tested against various 'non-strategic' missile targets including the re-entry vehicle of a modified SS-4 Sandal intermediate-range missile.[22] The conclusion may thus be drawn that Soviet ATBMs are intended to intercept not only the so-called tactical missiles Lance, Pluton and Pershing IA but even the medium-range Pershing II.

Moreover the Pentagon is convinced that the SA-12s are more suited than the SA-10s to intercepting some strategic missiles, especially SLBMs. This would mean that through further development of air defence systems mobile land-based BMD systems had come into being. Of course, the testing and deployment of multi-purpose air defence systems with a significant strategic ABM capability are prohibited under the ABM Treaty.

Delimitation of the operational options of Soviet ATBMs is far from simple, as important technical reference specifications are naturally not available. For example, it is not known what interception ranges can really be achieved or what warhead kill capability can be attributed to these systems. Moreover, the efficiency of the Soviet ATBMs will depend very heavily on the framework of defence installations in conjunction with which missile defence is carried out. It is of decisive importance that the supporting systems can guarantee earliest possible discovery and tracking of ballistic targets. It is certain that the Hen House class stationary radar unit, for example, will still have to be relied upon, as transportable or mobile radar units are not effective enough to pick up warheads at great distances. The Hen House class may, however, only pick up such longer-range missiles as Pershing IIs in time. As an alternative or complement, early-warning aircraft with infra-red sensors could be introduced. Powerful, airborne sensor platforms must be available particularly when short-range missiles such as the Lance are to be intercepted by an ATBM system.

Assessment

Bearing in mind the high technical demands and considerable operational constraints of ATBM defence, which after all also characterize Western development programmes, the prospects of success of the Soviet development efforts should not be overestimated.

The ATBM capability of the SA-12 should scarcely exceed that of the modified Patriot and even in combined operation with the SA-10 it may have only a limited capacity for self and point defence against intermediate- and short-range missiles. Against SLBMs and ICBMs its defence capability would certainly be even lower, leaving at best a marginal hard-point defence. The deployment of the first SA-12 Gladiator SAMs in the south-western part of the Soviet Union has already been interpreted in this way by the Pentagon, as SS-18 ICBMs are situated in the same area.[23] On top of this it is recognized that there is a possibility that the SA-12 could, in certain circumstances, be put into operation in combination with the SA-10 in order to defend mobile missile platforms (e.g., SS-25 ICBMs) against attack from SLBMs.

As the potential opponents of the Soviet Union can currently only use missiles with nuclear warheads, the Soviet ATBMs have to meet the much higher demands of 'warhead kill' right from the start. The 'mission kill' option, which could probably be carried out much more effectively with the same weapon systems, would only become relevant if, for example, NATO was one day in a position to field short-range ballistic missiles with conventional warheads in pursuing its follow-on forces concept. In anticipation of this challenge, it is quite possible that the Soviet Union already intends to create an ATBM component within the framework of a modified concept of integrated air defence.

Notes and references

1　Dr rer. nat. Hubert Feigl is a senior researcher at the Stiftung Wissenschaft und Politik Forschungsinstitut für Internationale Politik und Sicherheit, where he leads the programme group on new technology and international security. He has written extensively on space issues, see e.g., the papers cited in note 6.

2　Yost, D.E., 'Soviet missile defence and NATO', *Orbis*, vol. 29, no. 2 (Summer 1985), pp. 281-92.

3　Hoffman, F.S., *Ballistic Missile Defense and US National Security*, Summary report prepared for the Future Security Strategy Study, Oct. 1983.

4　Hassel, K.-U. v., 'The European Defense Initiative (EDI): Implications for Western security policy', *The European Defense Initiative—EDI—Some Implications and Consequences*, High Frontier Europa, Rotterdam, 1986.

5　Sorenson, D.S., 'Ballistic missile defense for Europe', *Comparative Strategy*, Fall 1985, pp. 159-78.

6　Feigl, H., 'Technische Herausforderungen und strategische Konsequenzen der amerikanischen Initiative zur Strategischen Verteidigung (SDI)', *Europa-*

Archiv, vol. 9, 1986, pp. 257-64; and 'Gegenmassnahmen als Herausforderung für SDI', *Europa-Arhiv*, vol. 1, 1987, pp. 11-22.

7 Wörner, M., 'A missile defense for NATO Europe', *Strategic Review*, Winter 1986, pp. 13-20.

8 Meyer, S.M., 'Soviet Theatre Nuclear Forces', Part I (Development of doctrine and objectives) and Part II (Capabilities and implications), *Adelphi Papers*, no. 187 and no. 188, IISS, London (Winter 1983-84).

9 Hines, K.L., 'Soviet short-range ballistic missiles—now a conventional deep-strike mission', *International Defense Review*, Dec. 1985, pp. 537-69.

10 Gormley, D., 'A new dimension to Soviet theatre strategy', *Orbis*, Fall 1985, pp. 537-69.

11 Nerlich, U., 'Missile defenses: strategic and tactical, *Survival*, May/June 1985), pp. 119-27.

12 Enders, T., 'Missile defence as part of an extended air defense', Konrad Audenauer Foundation, May 1986; and Allgaier, K.-H., 'Die Abwehr der Luftraumbedrohung Europas', *Wehrtechnik*, no. 7, 1986, pp. 38-41.

13 Enders, T., 'ATM: Europas Verteidigung auch gegen Flugkörper', *Europäische Wehrkunde*, no. 10, pp. 560-8

14 Greeley, B.M., Jr, 'Army missile intercept success spurs SDI theater defense study', *Aviation Week & Space Technology*, 29 Sept. 1986, pp. 22-3.

15 Wanstall, B., 'Neue Entwicklungen in der Luftverteidigung des Westens', *Interavia*, no. 12, 1986, pp. 1387-91.

16 Stevens, S., 'The Soviet BMD program', eds A.B. Carter and D.N. Schwartz, *Ballistic Missile Defense* (Brookings Institution: Washington, DC, 1984), pp. 182-220.

17 *Aviation Week & Space Technology*, 14 Nov. 1983, p. 23.

18 Rühle, M., Die strategische Verteidigung in Rüstung und Politik der UdSSR, *Berichte des Bundesinstituts für Otwissenschaftliche und Internationale Studien*, no. 40, 1985, Cologne; Yost, D.S., 'Soviet ballistic missile defense and NATO', *Orbis*, vol 29, no. 2 (summer 1985), pp. 281-92.

19 Department of Defense, *Soviet Military Power 1984* (US Government Printing Office: Washington, DC, 1984).

20 Department of Defense, *Soviet Military Power 1985* (US Government Printing Office: Washington, DC, 1985).

21 Rühle (note 17).

22 Longstreth, T.K. and Pike, J.E., 'US-Soviet programs threaten ABM Treaty', *Bulletin of the Atomic Scientists*, Apr. 1985, pp. 11-15.

23 *Jane's Defence Weekly*, 7 Mar. 1987, p. 359.

Paper 11. Impact on the two German states and their alliances

Elizabeth Pond

The Christian Science Monitor of Boston, Am Büchel 51c, D-53 Bonn 2, FRG

I. Introduction

Superpower decisions about ballistic missile defence (BMD) will have a different impact on the central front—the non-nuclear states of the Federal Republic of Germany and the German Democratic Republic—than on the nuclear weapon states France and the UK. Yet the consequences will be almost as profound. In broad terms the FRG and even the GDR perceive certain interests that parallel those of Britain and France.

These include: *(a)* preserving as much stability and leeway for political crisis management—and as much European input into crisis management—as possible; *(b)* maintaining adequate conventional defence, while keeping budget outlays within bounds; and *(c)* preserving as much of détente as possible.

In addition, there are two interests the Federal Republic shares with Britain and France but not with the GDR: *(a)* maintaining US nuclear deterrence of conventional war in Europe; and *(b)* sustaining commercial competitiveness during the dynamic changes of the third industrial revolution.

II. The Federal Republic and NATO

Until 1987, the Federal Republic's position on ballistic missile defence and the Anti-Ballistic Missile (ABM) Treaty was stated in such low-key fashion that it is easiest to perceive this position by examining incremental development rather than present-day exposition. A review of positions over the past four years is therefore necessary to convey fully the West German reservations about BMD.

When the subject first arose as a policy issue in the form of the US Strategic Defense Initiative (SDI), Bonn's centre-right government established certain guide-lines for its response. These have continued basically unchanged to the present day. They include the precepts that BMD developments should enhance rather than undermine strategic stability, should observe the restraints of the 1972 ABM Treaty, should avoid separating West European from US security, and should not deny NATO the option of 'flexible response' to any attack through use of either conventional or nuclear weapons.

All four points matched British and French interests, but the last was a particularly acute concern for a front-line non-nuclear state. As FR Germany read the conventional balance, the Soviet bloc enjoyed significant superiority over NATO on the central front in combat man-power and especially in heavy weapons. To offset this and to maintain deterrence Bonn had counted on the US threat of massive nuclear retaliation in the 1950s and on the West's tactical nuclear advantage in the European theatre in the 1960s and 1970s. After the Soviet buildup of SS-20s in the late 1970s and early 1980s, however, theatre nuclear superiority shifted to the Soviet side, and this began to curtail NATO's choices on defence. Now some West Germans feared that strategic defence might curtail choices even more: if it worked fantastically well, by creating a two-class system of security with superpower sanctuaries light years away from the European battlefield, and if it did not work well, by robbing conventional funds to pay the exorbitant costs of strategic defence.

All these considerations fed into the first official reaction after the Reagan Administration moved into a massive SDI programme in early 1984. At a NATO Nuclear Planning Group meeting in Cesme, Turkey, in April, Defence Minister Manfred Wörner worried aloud that strategic defence could introduce instabilities into the overall strategic balance and decouple West European security from that of the US. He also signalled his assumption that any deployments of strategic defences would not long remain a monopoly of the US, but would be duplicated within a few years by the USSR— and that Europe's calculation of its own interests must take this into account.[2]

Once he returned to Bonn, Wörner was quickly reined in. The West German Government had no wish to pick a gratuitous fight with the Reagan Administration over its still nebulous plans. And there were a few SDI enthusiasts in the Defence Ministry and in the conservative parties who wanted to go even further than benign neutrality on the subject and to give strong support for this new undertaking by NATO's senior ally. These officials hoped that through SDI the US technological genius might at last reverse the nuclear era's absolute vulnerability and restore to NATO's nuclear guarantor the invulnerability it had once enjoyed back in the pre-nuclear era. If this could be accomplished, they reasoned, it would lay to rest the recurring doubts about whether the US, in an emergency, really would risk Chicago to save Hamburg. If Chicago were defendable, then the US should have no inhibitions about committing nuclear forces to protect Hamburg.

At their most optimistic the SDI proponents argued that in a period of such technological dynamism the US might even re-establish the nuclear superiority it had wielded in the decades immediately following World War II; even after the Soviet Union caught up with the US in rudimentary SDI deployments, US technological exuberance might maintain a significant running lead.

In a distinction that would affect West German perceptions after the Reykjavik summit, however, these West German SDI adherents did not share one strong motivation of US SDI promoters. Steeped in the European historical sense of limited power and chronic inconclusive rivalry between nations, they tended to view maximalist aims as inherently unattainable. Instinctively they did not partake of the activist US expectation that problems can and should be solved once and for all. They believed that SDI might mitigate the hostile relationship with the USSR, but they did not really believe it could 'solve' it.

West German hard-liners did perhaps share the view of US hard-liners that Soviet assertiveness abroad was so intimately bound up with the domestic system that only a transformation of the whole society could ever moderate Moscow's foreign drive. But they were far less sanguine than their US counterparts that SDI might well force an all-out defensive and offensive arms race on the USSR that could push a state with only half of the United States' wealth into financial and economic ruin. It might well be true that the Soviet military giant had economic feet of clay, but Germans whose fathers had been repelled by ill-fed, ill-armed muzhiks in Stalingrad assumed that those clay feet could stagger on for a long time to come—and they saw a grave danger of rash action by a Soviet Union that felt itself being goaded into decline in a period of heightened confrontation.

After some internal debate Wörner's original misgivings carried the day (although he ceased to be their main proponent). His points were codified, in more diplomatic language, by Chancellor Helmut Kohl at the Wehrkunde defence seminar in Munich in February 1985, in the

Government statement of March 1985, and in the Chancellor's speech on the subject to the Bundestag in April 1985.

These declarations pointedly approved the research goals of the US SDI programme (omitting any mention of testing); urged that superpower research be 'channeled into cooperative solutions' and that the ABM Treaty be 'strengthened so long as no other common agreements are reached'; asserted that flexible response must be maintained 'so long as there is no effective alternative for prevention of war'; and presumed on-going US consultation with allies on these issues. In his Bundestag speech Dr Kohl added that instabilities 'must be avoided' in 'a possible transition phase from offensive nuclear retaliation to any stability based more on defence' and warned that 'the generation of new threats below the nuclear level must be avoided'.[3]

FR Germany did not trumpet the conditions it attached to its approval of SDI research. No immediate interests were at stake, and Bonn saw no sense in wasting political capital in an abstract debate with the popular US President over his pet project. Real decisions lay in the future in any case, possibly under a Reagan successor, and since Congress was conspicuously passive on the topic, Bonn had no natural allies in Washington working to establish political restraints on SDI. As a non-nuclear power, a Johnny-come-lately ally not enjoying the long special relationship of Britain to the US, and a diffident policy initiator in any case because of the shadow still cast by Hitler's Germany, the Bonn government chose to play its views in a low key.

The most low key of all were the senior Bundeswehr commanders. General-Inspector Wolfgang Altenburg expressed his opinions only in two guarded press interviews towards the end of his incumbency. In these he termed the US decision on SDI 'irreversible' but questioned its wisdom. Addressing the strategic issues, he emphasized the importance of keeping deterrence credible by sharing risk and avoiding any posture that would make a limited regional war look possible. He expressed concern that the dynamics of a race for strategic defence could tempt the superpowers to go for a destabilizing first-strike capability. And he warned that funds that were urgently needed for conventional defence could all too easily be diverted to BMD appropriations.[4]

Privately, other officials elaborated on the vexing question of costs. For planners already hard put to scrape up enough money for conventional forces out of basically static defence budgets in an era of rising personnel and modernization costs, it was hard to see what purse SDI allocations would come from other than that earmarked for NATO conventional force improvement. This, they feared, would leave NATO with the worst of both worlds, with, on the one hand, a partial SDI that was not good enough to guarantee security but was provocative enough to stimulate an unstable arms race, and, on the other hand, with growing deficiencies in conventional defence.

For all its reticence, the West German government cheered as British Prime Minister Margaret Thatcher visited Washington in December 1984 and persuaded Reagan to agree that SDI deployments would be negotiated 'in view of treaty obligations' and be designed 'to enhance, and not to undermine deterrence'. It cheered again when British Foreign Minister Sir Geoffrey Howe publicly warned in March 1985 against putting faith in a chimerical Maginot Line of SDI. And after the Reagan Administration revised the 13-year-old traditional understanding of the ABM Treaty to a permissive 'broad interpretation' in October 1985, the West Germans joined other Europeans in welcoming the Administration's corrective statement that 'policy' would continue to follow the narrow interpretation.[5] Bonn appreciated further the pledge to NATO ministers in Brussels in November 1985 that Washington would consult with its allies before again revising this policy.[6]

Having resolved its own strategic position on SDI by spring 1985, the West German Government welcomed a diversion of its domestic strategic-defence debate at that point to a peripheral issue: access by West German firms to US SDI contracts and high-tech spin-off. For different reasons Chancellor Kohl, the Foreign Ministry and business-men were all interested in getting US-West German agreement on the subject.

For Kohl, whose first order of foreign policy was to maintain cor-dial relations with the US President, the agreement that was finally signed in spring 1986 was a useful vehicle that allowed Washington to claim West German political support for SDI without compromising Bonn's strategic reservations. For the Foreign Ministry, the commercial issue was a red herring that could side-track the real trans-Atlantic security debate until the maturing of SDI technology and US politics brought some realism to the more extravagant claims advanced for SDI. For West German industry the issue was a useful lever for establishing some general ground rules for West German companies' commercial use of technology developed under US contracts. The business world, with the exception of a few aerospace giants, was sceptical about the amount of dollar contracts it would get—and the research ministry was downright cynical, openly preferring the new pan-European Eureka programme as a seed-bed for new technologies. But on balance it seemed better to use the opportunity than not to use it.

In the end the contracts were in fact disappointing (as were the rather anodyne ground rules on technology access), and a protectionist Congress balked at increasing the orders to Europe. As US budget deficits, foreign debts and trade gaps plunged the dollar to record lows, a further consensus coalesced in the West German business community that the US military promotion of industry was actually harming the economy and accelerating the competitive decline of the United States. Japan's decidedly civilian high-tech orientation came to be seen as a much more efficient motor of progress. All this meant that no powerful defence-industry lobby for SDI arose in FR Germany comparable to that in the US.

Thus far the West German response to SDI largely paralleled the British and French reaction, if with increased emphasis on the issue of extended deterrence and the nuclear impact on the conventional balance. But in one particular Bonn has differed notably from London and Paris: anti-tactical ballistic missile (ATBM) defences. Defense Minister Wörner quickly became, along with NATO European commander Bernard Rogers, the main prophet of developing a tactical ABM system (called in Bonn, largely for public-relations purposes, 'extended air defence').

In part, Wörner drew the conclusion of inevitability from SDI; if strategic missile defence was being developed, then a tactical missile defence was the next logical step and might be needed not only against a nuclear attack, but also against any future threat from conventional missiles of the Warsaw Pact. In part, Wörner counted on ATBM defences to spur development of other high-tech improvements in conventional weapons that might redress the existing conventional imbalance in Europe.

Possible circumvention of the ABM Treaty through ATBM development was no real issue in the West German discussion. The SDI linkage that occupied West Germans was instead a technical one: would the more mature (basically terminal-phase) technologies that had been developed for missile tracking and interception since the 1970s mean that ATBM systems against the shorter and lower trajectories of shorter-range missiles could be deployed earlier than SDI? Or would ATBM effectiveness against the short flight times have to await the greater sophistication of more exotic technologies in the parent SDI programme? West German opinions differed, reflecting the mixed opinions among US SDI experts. But Bonn thought it worthwhile to explore the possibilities .

Britain and France displayed a notable lack of interest in the whole enterprise. They saw no harm in going along with FR Germany in funding preliminary ATBM studies in NATO, but they were highly sceptical about the opportunity costs of sinking vast sums into any more substantive development programmes at the possible sacrifice of more offensive capability in British and French nuclear missiles or more West German heavy conventional weapons.

For its part, the US welcomed West German interest in pursuing ATBM defence as demonstrating European engagement and therefore political endorsement of SDI. Defense Secretary Caspar Weinberger saw to it that the SDI programme conspicuously solicited European contractors for ATBM architecture studies.

There things stood until the Reykjavik summit.

III. The Reykjavik summit

Two aspects of the superpower summit at Reykjavik in October 1986 shocked West Germans as they did other European allies of the US.

The first was Reagan's casual willingness to dispense with all ballistic missiles 10 years hence—and to advance this goal without having commissioned any prior studies by NATO or even by the US Joint Chiefs of Staff to explore the consequences for deterrence and war fighting. The second was Reagan's readiness to set this target with the leader of the other superpower over the heads of US allies. (There was a third shock as well, as the Soviet Union provisionally accepted the four-year-old NATO proposal to eliminate all INF missiles in Europe, but this involved self-reproach for having made the offer in the first place and did not concern either intra-alliance relations or SDI.)

Thus, it was all well and good for US hard-liners to argue that getting rid of all ballistic missiles would be a boon at the strategic level, since this would remove a Soviet advantage in such areas as heavy ICBMs with a theoretical first-strike capability against the United States, while leaving untouched the US advantage in aircraft and air-launched and sea-launched cruise missiles. This might conceivably guarantee US invulnerability and strategic deterrence.

In the European view, however, it certainly did not guarantee extended deterrence in Europe. Quite apart from pulling contracted

Tridents out from under the British Government, such a regime would mean that precisely those NATO weapons with guaranteed penetrability would vanish and would leave NATO aircraft (and sea-launched cruise missiles, since all land-based cruises would be removed under the INF deal), facing the densest and most mobile air defence in the world. On the ground it would reduce NATO's nuclear defence to artillery and other tactical war-fighting weapons that promised to devastate Germany and would, the Germans thought, erode the credibility of flexible response.

This in turn could make Europe safe for conventional war, or for political extortion based on the conventional superiority of the Warsaw Pact.

Intellectually, it would be possible to make the case that these prospects were not all that new and should not have caused such a shock. Doubts about the credibility of nuclear extended deterrence had been around since Charles de Gaulle's time, and NATO's assessment of Soviet buildup of nuclear superiority in the European theatre in the late 1970s and early 1980s certainly called into question the military utility of any NATO escalation from conventional to nuclear warfare. As analysts became increasingly convinced of the non-utility of nuclear weapons except for deterring use of the adversary's nuclear weapons, deterrence by all-but-certain nuclear retaliation for any successful conventional invasion was already shading into deterrence through an adversary's uncertainty about irrational but possible nuclear escalation.

Intellect was one thing, however, viscera another. West Germans felt Reykjavik in the solar plexus—and those who felt the most aggrieved were the SDI enthusiasts who considered themselves the USA's best friends. The point that had long been an axiom to the sceptics suddenly became an axiom to the enthusiasts too: that the corollary of a world of strategic defence was a world without usable ballistic missiles—and therefore, however protected Chicago might be, a world with tenuous extended deterrence. In such a Europe the conventional superiority of the Warsaw Pact would gain considerable political weight, and the Federal Republic could be reduced to the second-class security it feared.

Moreover, the fact that the US had set this goal with the other superpower without even discussing it previously with NATO allies made the realization particularly painful. The whole achievement of nuclear consultation so assiduously cultivated in NATO since the INF decision of 1979 seemed to have evaporated.

Chancellor Kohl spoke of these concerns when he visited Washington immediately after Reykjavik. But once again the West Germans deferred to the British for the clearest articulation of European misgivings. Prime Minister Margaret Thatcher spoke for all the European allies in establishing in the Camp David statement signed jointly with President Ronald Reagan in November that priority should go to the more orthodox arms-control goal of 50 per cent cuts in offensive strategic warheads (omitting mention of banning ballistic missiles) and repeating 'that SDI research should continue within the terms of the Anti-Ballistic Missile Treaty'.

As the 1987 ABM Treaty review year opened and opponents of the Treaty inside the Reagan Administration lobbied the President to deploy a rudimentary SDI system by the early 1990s—or at least to test SDI under the broad interpretation—in a way that would effectively scrap the Treaty, West German officials made clear to diplomats and journalists their interest in adherence to the Treaty in general and the narrow interpretation of the Treaty in particular. And Kohl himself made one of the strongest public statements of any European leader in saying that interpretation of the Treaty was a matter for both signatories to work out 'in co-operative solutions' and not for unilateral change.[7]

Foreign Minister Hans-Dietrich Genscher also set the tone of the Government's view in publicly urging conciliation rather than confrontation in Western attitudes toward a Soviet leader who was trying to open up his society and overhaul his laggard economy. He rebutted too the rationale both for SDI deployment that would violate the ABM Treaty and for a broad interpretation of the Treaty in saying that it would be 'dangerous'

...to fall prey to the illusion that the Soviet Union was acting from a position of weakness which must be exploited or even aggravated. Firmness is called for, but a policy of projecting strength, of striving for superiority, of checkmating the other side by outarming it, must be consigned once and for all to the past as an outmoded approach...in the West as well as the East; such an approach would inevitably lead mankind to catastrophe.[8]

In this the West German Government perceived its interests as going beyond purely military issues of SDI, which it had assumed would sort themselves out with time on their own technological merits. In the US hard-liners' bid for early and, the West Germans believed, technically premature deployment of SDI, officials feared a deliberate attempt to sabotage arms control until such time as the superpowers were locked into an all-out defensive and offensive arms race. That kind of instability even the early West German SDI fans could see as only harming their interests in maintaining détente with the German Democratic Republic and in nurturing their own public security consensus after the divisive debate of the early 1980s about INF deployment.

To be sure, SDI would not require any deployments on West German soil, and public *angst* and the strong anti-Euromissile peace movement of the early 1980s had faded ever since the INF stationing actually went ahead in 1983 and East-West relations turned out to be none the worse for it. But government officials were wary of the swiftness with which the politically attentive public, now sensitized to security issues, might re-engage in the next controversy over defence.

Moreover, the psychology of living on the fault-line between East and West meant that West Germans were less ready than Americans to accept the thesis that nuclear deterrence is inherently so stable that East and West can proceed by an instinctive sense of restraint and do not need to maximize predictability through written arms control agreements.

The net result was an unusual unity among West German, British, French and other European members of NATO in seeing SDI *research* as beneficial in prompting a new Soviet appreciation of the merits of arms control and strategic stability but SDI *testing* as harmful at the point at which it could kill the ABM Treaty and therefore arms control.

As of this writing the common European perception is not leading to trans-Atlantic crisis. Instead, it is entering the factional infighting in Washington to reinforce moderates within the Reagan Administration (and Democratic sceptics on strategic defence in Congress). Barring a decisive victory by hard-liners in Washington, this would seem to be Europe's, and West Germany's, likely continued role in the BMD issue.

IV. GDR and the Warsaw alliance

Far less evidence is available about East German than about West German attitudes toward BMD. Virtually the only clue is provided (by analogy) by the enthusiasm with which the GDR embraced the peace issue in 1983/84 and continued to do so after this policy no longer held any instrumental value for a Soviet leadership that was by then stressing a sense of crisis in East-West relations. The East German reactions at that time suggest that in broad terms the GDR finds controlled détente more compatible than heightened East-West confrontation, and that it balks at moves that aggravate tensions and might require increased East German military outlays.

There is, of course, as much foreboding in the GDR as in the FRG that Germany is the likely battlefield in any war. And there is the normal echo of Soviet criticism of the US SDI programme in the East German press. This represents—there is every reason to believe—the real opinion of the East German leadership, since SDI is seen as a programme that could destroy arms control and thereby détente and the special German-German relationship that détente allows.

There is no equivalent discussion in the East German press of Soviet efforts in strategic defence or speculation about their impact on Moscow's East European allies. Quite apart from the greater constraints on public airing of Soviet-East German relations than on discussion of US-West German relations, nuclear questions simply do not have the same urgency inside the Warsaw Pact as they do within NATO. Since military and political relationships among member states are predicated much more on conventional than on nuclear forces, there is no issue of the superpower ally's nuclear umbrella or 'extended deterrence' comparable to that within NATO. Nor is there any junior Soviet ally like Britain or France with its own independent nuclear weapons.

The closest an outside observer can come, therefore, to discerning East German views on the potential impact of BMD on the Warsaw alliance is to look at the East German treatment of theatre nuclear questions in 1984.

In November 1983 the Federal Republic, as the key NATO ally to station intermediate nuclear forces on its territory, made it clear that it would proceed with deployment. The first Pershing IIs went into place and were operational by the end of the year.

Within a few months Soviet and West German relations deteriorated—but East and West German relations conspicuously did not. GDR party leader and Chairman of the State Council Erich Honecker, who before the stationing had warned of an 'ice age' in East and West German relations should the NATO deployment go ahead, dropped all talk of this as soon as the Pershing IIs actually arrived. Instead, he spoke of the need to 'limit damage' and pursue the 'coalition of reason' between the two German states—however bad superpower relations might be—in order to prevent war from ever again starting on German territory. He also began musing about a special role for medium-sized states in international relations. Indications were that this policy met with popular approval in the GDR, and that this approval was important to Honecker. Then as the Soviet media revived old pre-détente charges that West German 'revanchists' wanted to regain former German territory now in Poland, the GDR soft-pedalled these accusations. And Honecker continued planning for his maiden visit to FR Germany, set for the fall.

In the same period, most unusually, the East German press carried some readers' letters objecting to the increased military costs that East German citizens would have to bear because of the new Soviet short-range missiles being stationed in the GDR in an advertised response to the new NATO deployments.

East German-Soviet differences led to an unprecedented public argument in the summer of 1984, not about weapons, but about inter-German cosiness. *Pravda* published a thinly veiled criticism of the GDR for putting good relations with the Federal Republic above solidarity with the GDR's senior ally and the Soviet attempt to freeze out FR Germany diplomatically for approving the NATO deployments. The Federal Republic was trying to undermine the GDR's sovereignty and its socialist system, *Pravda* asserted, and in any case (good) East and West German relations could not be separated from (bad) super-power relations.[9] In the wake of this exchange Honecker called off his visit to the Federal Republic and was still waiting to reschedule it three years later.

There is no evidence of East German lobbying of Moscow—comparable to West German lobbying of Washington—to adopt particular positions on strategic defence. There is evidence of East German lobbying, however, for a separation of INF issues from SDI and for removal of all theatre nuclear weapons (including those of short range) from the GDR in officials' repeated references to INF as 'the devil's tool'. And the history of 1984 suggests strong East German interest in restraining BMD development and achieving a superpower arms control agreement that could avert an intensified arms race and heavier military expenditures for the GDR as well as for the Soviet Union.

More fundamentally, this history suggests East German interest in avoiding the East-West tension that would accompany a new arms race and disturb the relaxed inter-German relationship that has now become a staple of the GDR's foreign policy.

Notes and references

1   Elizabeth Pond is the Bonn Correspondent of *The Christian Science Monitor*.

2   Taylor, P., Reuters, 'NATO ministers informed but not convinced on Star Wars', Cesme, Turkey, 4 Apr. 1984; 'Wörner besorgt über neues US Abwehrsystem', *Hannoversche Allgemeine Zeitung*, 7 Apr. 1984.

3   Stellungnahme der Bundesregierung zur Strategischen Verteidigungsinitiative (SDI) des Präsidenten der Vereinigten Staaten von Amerika, *Bulletin des Presse- und Informationsamtes der Bundesregierung*, no. 35, 29 Mar. 1985; 'Erklärung der Bundesregierung zur Strategischen Defensivinitiative des Präsidenten der Vereinigten Staaten von Amerika, *Bulletin des Press- und Informationsamtes der Bundesregierung*, no. 40, 19 Apr. 1985.

4   Feldmeyer, K., 'Die Risikogemeinschaft darf nicht zerstört werden, Ich habe grosse Sorge wegen SDI', *Frankfurter Allgemeine Zeitung*, 29 Nov. 1985; 'Krieg in Europa wird wieder möglich', *Der Spiegel*, 10 Feb. 1986.

5   Interview with Foreign Minister Hans-Dietrich Genscher on Süddeutsche Rundfunk in the programme Politik der Wiche, 12 Oct. 1985.

6   Interview with Hans-Dietrich Genscher on Norddeutsche Rundfunk, 15 Oct. 1985.

7   Interview, *Neue Osnabrücker Zeitung*, 25 Feb. 1987.

8   Speech in Davos, Switzerland, 1 Feb. 1987: *Mitteilung für die Presse*, no. 1022e/78, Foreign Ministry, 4 Feb. 1987.

9   Bezyminsky, L., 'In the shadow of American missiles', *Pravda*, 27 July 1984; this and other articles from the exchange are available in English in Asmus, R.D., 'East Berlin and Moscow: The documentation of a dispute', *Radio Free Europe Occasional Papers*, no. 1, 1985.

Paper 12. 'For the benefit and in the interests of all': superpower space programmes and the interests of third states

Donald L. Hafner[1]
Department of Political Science, Boston College, Chestnut Hill, Mass. 02167, USA

I. Introduction

Like most states the superpowers often believe that their own problems and concerns are far more consequential than those of other nations, and that the good they do for themselves is really for the benefit of all. Sometimes their beliefs may be correct. If the hazards of nuclear war between the superpowers are truly as great as some profess, and if traditional policies of deterrence are as unstable as some fear, then perhaps the superpowers would be serving the interests of all mankind by abandoning the ABM Treaty and seeking ways to make nuclear weapons 'impotent and obsolete'.

But other states have their doubts. Success is by no means assured in the quest to replace offensive weapons with defensive systems, and there is broad agreement that a world of imperfect defences could be far more hazardous and unstable than the one we now live in. President Reagan conceded these points in his March 1983 speech; General Secretary Gorbachev has repeatedly affirmed them.

Where success and benefit for all are not assured, other states will certainly ask what harm may be done to their interests as the superpowers develop and deploy strategic defences. And each superpower should certainly ask what complications may arise for its defence programme if other states act to protect their own interests. What will a world of strategic defence be like? Most often when this question is asked, discussion turns to the ABM Treaty. However, the ABM accord is not the only treaty of importance here. This paper examines the potential consequences of superpower defence programmes in a realm that both superpowers pledged by treaty in 1967 to explore and use 'for the benefit and in the interests of all countries'—outer space.

II. Third party interests in outer space

In the 20 years since the superpowers made that pledge in the Outer Space Treaty, the number of other states exploiting outer space has increased greatly. In 1967 France was the only state other than the superpowers to have successfully launched a satellite into orbit; by 1987 16 nations (China, India, Japan, and the 13 members of ESA, the European Space Agency) had acquired the capacity to orbit satellites on their own boosters. And by 1987, satellites belonging to the Arab League, Australia, Brazil, Canada, China, Czechoslovakia, ESA, France, India, Indonesia, Italy, Japan, Mexico, The Netherlands, Spain and FR Germany had circled the globe, performing a host of functions ranging from general scientific exploration to communications, meteorology, remote sensing and military reconnaissance. In 1964 the International Telecommunications Satellite Organization (INTELSAT) was created by 11 member states, to establish a geosynchronous satellite communication system; by 1987 INTELSAT had 109 member states and carried more than two-thirds of the world's international phone communication for 145 user nations, as well as most live television coverage. In 1979 the International Maritime Satellite Organization (INMARSAT) was established to provide a satellite communication network for ships, and eventually for commercial aviation; by 1987 INMARSAT had 43 member states and served over 4000 ships.

The willingness of the superpowers to share some of their own space programmes has also fostered growing involvement of third parties in outer space. Astronauts and cosmonauts of more than 10 nations have participated in the manned space programmes of the Soviet Union and the United States. Data from superpower meteorological satellites (and those of other states) are transferred to the World Meteorological Organization (WMO), to be shared with all nations. In addition, more than 1000 readout stations in 125 nations can receive weather data and global weather images from these satellites directly. Data from the US remote-sensing satellite, LANDSAT—useful for assessing agricultural conditions, mapping water supplies and locating other natural resources—is available for purchase by any state or person. Ground stations in 10 other nations are licensed to receive LANDSAT data directly and distribute it to domestic and foreign customers. The superpowers, along with six other states, have established a co-operative Search and Rescue Satellite System (COSPAS/SARSAT) able to detect and locate distress signals from special transmitters carried by thousands of ships and hundreds of thousands of aircraft. Since the satellites that provide such services typically cost $30–60 million each to place in orbit, they represent a resource that would be too costly for most third party users.

As the international use of satellites has grown, so has the international governance of outer space. In 1958 the UN General Assembly formed the *ad hoc* Committee on the Peaceful Uses of Outer Space (COPUOS), with 11 members; by 1987 COPUOS was a permanent UN committee with 53 member states. A substantial part of international law and co-operative activity regarding outer space has had its origins in COPUOS, including the Outer Space Treaty. In January 1967, when the Outer Space Treaty was opened for signature, there were 65 signatories; by the end of 1967 another 20 states had signed; since the Treaty went into force more than 20 additional states have acceded to it. The International Telecommunications Union, which co-ordinates geostationary positions and radio frequencies of satellites, has even wider representation, with more than 150 member states. The World Meteorological Organization, a co-operative planning and co-ordinating body, is open to membership by any state with weather services. The members of INTELSAT and INMARSAT share responsibilities and governance under their own charters.

There is hardly a state on the globe, then, that does not benefit in one way or another from services provided by space satellites. At the same time, the international network of legal and co-operative organizations with responsibilities over space activities has become so far-reaching and complex that there are few states without at least some voice in the governance of the peaceful uses of outer space.

III. The challenges to third party users

The superpowers no doubt believe that they are quite respectful of the interest all states have in international access to space. Granted, from time to time one or the other superpower has protested—or even made verbal threats against—some activity in space, be it photosurveillance, direct broadcasting from satellites or public release of data from remote sensing satellites. Yet in practice, both have tolerated open access to space by all.[2]

Now all this may change. Open access by third parties may soon confront challenges by three trends in superpower space programmes: the growing reliance on satellites for military support functions; the development of sophisticated anti-satellite weapons (ASATs); and the apparent commitment to space-based ballistic missile defence (BMD) and ASAT countermeasures to BMD. If these three trends continue, the superpowers may—indeed, logically they *must*—impose new restraints on access to outer space.

As we turn to the reasons and possible character of these restraints on access, two points must be kept in mind. First, any discussion of future space programmes is speculative and therefore must be approached with caution. It is extremely difficult to predict the direction of superpower ASAT and BMD programmes. These programmes are in flux, and vital decisions about their direction have not yet been made. The United States, for instance, began developing a very sophisticated ASAT interceptor in 1976, reportedly with the intent of deploying more than 100 interceptors at two sites no later than 1987. Yet 10 years later, the weapon had been tested only once against a satellite target, Congress had cut funding and imposed a test moratorium, and the US Air Force reportedly wished to abandon the device altogether and develop an even more sophisticated weapon.[3] Similarly, a recent study of space-based defences published by members of the Soviet Academy of Sciences mentions a variety of countermeasures to defences, including 'space mines' (satellites filled with explosives), ground-based lasers, swarms of small pellets placed in orbit and nuclear blasts set off in the upper atmosphere.[4] Yet whether either side would actually develop and test such weapons would presumably depend upon the success of BMD research programmes, something that at the moment cannot be known.

A second point to keep in mind is that it may not be easy for specific third parties to decide whether on balance they will be better or worse off as the superpowers proceed with space programmes. The inconveniences imposed by the superpowers will not rest evenly upon all third parties, and some may actually derive benefits from superpower BMD programmes. Greatly reduced launching costs, highly advanced remote sensing devices, high data-rate communications, exotic new satellite construction techniques and materials, all these might be by-products of superpower space programmes, even if space-based BMD ultimately proves unfeasible. In many respects, these by-products might have greater significance for third states: if launch costs could be reduced from the current $2000-$7000 per kg to $225 (a goal of the US

SDI programme), even states with meagre resources might enjoy what once was reserved for the rich—their own satellites, devoted entirely to their own needs. Other states, bound by political association or military alliance with the superpowers, may feel that their security is enhanced by superpower ASAT and BMD programmes. This is an issue, therefore, where the interests of superpowers are not diametrically opposed to the interests of *all* third parties. In this realm as in all others, each state must assess the balance of benefits and burdens carefully.

Keeping both the speculative nature of this discussion and the importance of weighing both benefits as well as costs in mind, we can turn now to a more detailed consideration of the ways in which superpower ASAT and BMD programmes conceivably may interfere with third party use of outer space.

IV. Space weapon testing

The earliest problems for third party users of outer space will stem from tests of space weapons, beginning perhaps decades before large-scale deployment. And the earliest of space weapons—indeed the only ones so far—are the superpower ASATs.

ASAT testing

Until now, the prime justification for ASATs was the burgeoning use of satellites by conventional and strategic military forces for surveillance, navigation, communication, attack early warning, and so on. In the future, the most attractive targets for ASATs may be space-based BMD weapons. As the reasons for ASATs multiply, so too will the number of ASAT tests.

· ASATs can take a variety forms: manoeuvring interceptors placed in orbit or launched directly from the ground; space mines; high-powered lasers (ground- or space-based); and so forth. But whatever the mode of attack, during tests the ASAT will create an area around the test target that would be lethal or damaging to other space vehicles passing through it.[5] In addition, the test target, if destroyed, will create an expanding volume of debris that may also be hazardous. The ASATs themselves may inject debris, waste fuel or gases, spent boosters, and so on, into orbit; if the ASAT destroys itself in the course of the test, or is deliberately destroyed as a security precaution, the amount of hazardous debris will be larger.

Thus far, superpower ASAT tests have been infrequent occurrences, yet the debris hazard they cause is already a serious problem. The Soviet Union has tested its orbiting ASAT interceptor only 20 times since 1968, and never more than four times per year. The United States originally planned only 10 tests of its ASAT against satellite targets over a period of four years, before deploying the weapon. In contrast, the two superpowers have conducted over a hundred launches of satellites per year, on average, since the mid-1960s.[6] Nonetheless, of 6000 orbiting objects trackable by radar and reported by NASA in mid-1986 (that is, objects larger than about 10 cm), roughly 900 were generated by ASAT tests. Of 98 radar-trackable fragments generated by the first Soviet ASAT test almost 20 years ago, 60 pieces are still in orbit. The single US ASAT test against a target in September 1986 produced over 200 trackable fragments.

Put differently, during the past decades when ASAT weapons were not being vigorously developed or tested, superpower ASAT tests constituted less than 1 per cent of all space launches, yet they contributed 15 per cent of all current trackable debris.[7] As the superpowers increase the pace and variety of ASAT tests, it will be impossible to ignore these hazards to their own space programmes and to those of other states.

BMD Testing

In many respects, superpower testing of BMD weapons in space will pose the same problems for third parties as ASAT testing. The precise form of these weapons is still unknown—they may be high-powered lasers or kinetic energy warheads propelled by any of several means. Whatever their form, however, they will probably be tested first in an anti-satellite mode, because initial tests against orbiting instrumented test targets will provide a better assessment of performance, at lower cost, than tests against ground-launched ballistic missile warheads. In the beginning, perhaps no more than one target will be engaged per test, and the debris, waste gases, spent warhead casings, and so on, produced may be roughly the same as in an ASAT test. ASAT testing itself will be continued, to cope with a major challenge for any space-based BMD weapon—defending itself against 'space mines' and other orbiting BMD weapons.

If development goes further, however, prototype BMD weapons would have to be tested in modes resembling operational performance, with hundreds of simulated engagements, all within minutes of each other. If the test targets are boosters or ballistic warheads, most of their debris would shortly re-enter the atmosphere and so be less of a hazard than in ASAT tests. On the other hand, an effective space-based BMD weapon will have to be tested out to ranges of 1000 km or more, perhaps in a network of several weapons all tested simultaneously. The region of space occupied by tests of this sort would be significantly larger than for ASATs.[8]

What is the solution? Left to their own preferences, the superpowers would probably fall back on precedent, on the practices they follow when conducting military exercises and missile tests in international waters. Prior to exercises or tests, advance notice is given to mariners and airmen that designated ocean areas may be hazardous. Strictly speaking, it is unlawful to ban access to international waters during these periods of notice, but the practical effect is the same, and the testing state often makes vigorous efforts to intercept ships or aircraft that may innocently or deliberately intrude during a test. This precedent has special appeal for the superpowers because these practices have been adopted as well by third party states that conduct military and civil missile launches over international waters.[9]

There are crucial differences, however, between tests on the high seas and tests in outer space. For one, it is inherently easier for a ship or aircraft to avoid a restricted region of ocean. By design, all ships and airplanes are equipped for manoeuvring, with fuel, motors, guidance and human operators to make prompt decisions. If a hazard exists tens or hundreds of kilometres ahead, they can get out of harm's way if given a few minutes warning. In contrast, most satellites are not equipped for altering their orbital characteristics by tens or hundreds of kilometres, even with ample warning. By the laws of orbital mechanics, such satellites will be carried inexorably toward hazards that may originally have been thousands of kilometres away on their orbital paths.

A second difference is that in contrast to ships and aircraft, satellites are susceptible to damage by the smallest bit of debris or momentary illumination by powerful lasers, because satellites have fragile, lightweight construction, high orbital velocities, and vulnerable sensors.[10] Finally, the by-products of space weapons tests may persist long after the actual test and spread considerably beyond the test region, so that in effect a whole orbital band may become hazardous for some time—an area of permanently restricted access.

These problems deserve attention. Outer space is a vast realm, and while the probability of collision between satellites and debris is not large, neither is it trivial.[11] The problem has already become burdensome to the superpowers. The North American Aerospace Defense Command (NORAD) devotes a good deal of effort to tracking the US space shuttle and calculating all potential collisions with space objects each time the shuttle goes aloft. Special procedures for venting unburned fuel have been adopted by NASA to reduce the debris caused by spent boosters exploding in space. The American Institute of Aeronautics and Astronautics has proposed (thus far unsuccessfully), an international agreement among all space users to ban explosions of satellites or restrict them to low orbits, where debris is more rapidly swept from orbit by atmospheric friction.[12] Certainly the superpowers have their own incentives to be careful where and how they conduct ASAT tests, since they have a great number of their own satellites and manned vehicles in vulnerable orbits. The unpredictable side-effects on their own satellites, for instance, may prove the strongest deterrent to tests of nuclear-armed ASATs by the superpowers.

Many third parties would no doubt favour a general prohibition on ASAT testing. That may not be feasible, even if it were desirable.[13] Short of a general ban, third parties do have some channels for protecting their interests. Article IX of the Outer Space Treaty already obligates each member to consult with others if its space activities might cause 'potentially harmful interference' with the space activities of other states. Article III of the Convention on International Liability for Damage Caused by Space Objects, signed in 1972 by both superpowers, would entitle third parties to compensation if space objects belonging to a superpower were at fault in damaging a third party satellite. Concerns can also be pressed within the co-operative organizations regulating space activities.

But the truth must be stated bluntly: the superpowers have not surrendered significant authority to international bodies in which third parties hold the majority voice, and, in any case, many third parties can ill afford to irritate the superpowers upon whom they are dependent for satellite services. Nothing illustrates the problem better than the fact that the evidence a third party would need to prove its satellite had been damaged by debris from a weapon test is currently available from only one source—the space tracking systems of the superpowers themselves.[14] If these matters are to be resolved successfully, it must be by appeals to the common interests of superpowers and third parties alike.

V. Space weapon deployments

The greatest consequences for third party users of outer space will come with the actual deployment of space-based BMD by the superpowers. A highly unstable strategic situation could result if the defensive system of either side were vulnerable, or were believed to be vulnerable to direct attack. To ensure the survivability of its BMD systems, each superpower will have to regulate the region of space around its BMD satellites. That means that during peace-time, and certainly during crisis or conflict, third party users will have to adapt their space activities to what the superpowers find tolerable.

A full-fledged BMD system would probably have a large number of component parts, located at different altitudes. The actual battle station for boost-phase and mid-course intercepts would presumably be in low earth orbit. These might be kinetic energy launchers, lasers or 'fighting mirrors' for redirecting ground-based laser energy. The number required may be as few as a hundred, or more than several hundred, depending upon such factors as lethal range, rate of fire, or level of defence effectiveness desired. In somewhat higher orbits there might be several dozen satellites with sensors for detecting and tracking warheads during mid-course. Higher still, at geosynchronous orbit, might be a handful of early-warning satellites, and perhaps also vast mirrors for redirecting ground-based laser energy to 'fighting mirrors' at lower orbits. And higher yet, at supersynchronous orbits, might be a large number of communication and battle management satellites, linking the entire network together. Each authentic satellite might be accompanied by a number of decoys. Manned space stations at various altitudes might be needed for routine maintenance, with crews in transit on a regular basis to authentic satellites and to decoys. Both superpowers would presumably deploy the whole array of components.

The precise architectural details of a BMD system have been the subject of furious but inconclusive debate among experts. For third party users, the details may matter less than how good the BMD system is: the better it is, the fewer problems it may pose for third parties. In peace-time or in crisis, the principal concerns of each superpower will be to avoid collisions between its satellites and other objects, and to enforce keep-out zones against its opponent to prevent the opponent from placing space mines, other ASATs, or its own BMD battle stations within lethal range. The total volume of space encompassed within these keep-out zones will depend upon techniques applied to ensure the survivability of BMD systems against attack. A variety of active and passive techniques are conceivable, including armouring, shielding, decoys, proliferation of key satellites, counter-attacking the attacker and intercepting ASATs. The greater the effectiveness of such techniques, the smaller the keep-out zone would have to be around each satellite.

More to the point, if a superpower has truly devised a BMD system capable of detecting, discriminating and tracking objects as small as missile warheads 'from birth to death', it presumably could also track and discriminate third party satellites from the space mines and ASATs of its opponent. It could therefore allow free passage of third party satellites through keep-out zones, unless of course it feared the hostile intentions of a third party (e.g., if the third party were an ally of the opposing superpower). The only concern would be the risk of collisions, but again, an extremely good BMD satellite could presumably detect and evade approaching objects or, in the extreme, destroy them.[15]

The problems arise if BMD systems are *not* very good. If they cannot discriminate non-hostile objects, cannot manoeuvre or defend themselves instantly and readily, and cannot survive except with great numbers of redundant satellites, then large keep-out zones enforced against all parties will be necessary, in peace-time as well as in crisis. Entire orbital bands or regions, several hundred kilometres in depth, might be declared restricted zones, with passage through the zone allowed only in conformity with rules set down by each superpower. Proposals along these lines have already been put forward in the United States and the Soviet Union.[16] Poor space-based defences raise grave questions about strategic instabilities in any case; poor defences that must protect themselves by defending large keep-out zones only heighten the instabilities.[17]

We noted earlier that a practice of closing off areas of space temporarily during space weapons tests might find legal precedent in the law of the sea. In contrast, keep-out zones around deployed weapons in space would have to be permanent, and as such they would appear to be banned by Article II of the Outer Space Treaty, which states that space 'is not subject to national appropriation by claim of sovereignty, by means of use or occupation, or by any other means'. The superpowers could ask third party signatories to amend the Treaty. However, the amendment procedure in Article XV of the Treaty stipulates that amendments go into force only for those states accepting the amendment, once it is approved by a majority of signatories. It would defeat the whole purpose of keep-out zones if they could be enforced only against space objects known to belong to consenting states. While the superpowers would no doubt prefer to operate under the authority of an amended treaty, they may also be willing to go ahead without it.

VI. The ASAT capabilities of third parties

This brings us unavoidably to the question of whether, with the growing space capabilities of third parties, the superpowers might find third parties able to enforce their rights of access to outer space with ASAT weapons of their own.

The very fact that we have to ask the question suggests how far the whole issue of space weapons can take us from the original spirit of the Outer Space Treaty. Nonetheless, the question needs to be asked if we are to understand the politics that may govern future negotiations on rights of access in a world of space-based BMD weapons. The question is also worth asking because ASAT weapons are an issue in their own right, and we should understand the dimensions of this space weapons problem as well.

At first glance, the ASAT task seems deceptively simple, and occasionally assertions are made that quite ordinary space activities, such as orbiting and manoeuvring a satellite, would grant a crude ASAT capability.[18] Perhaps the simplest of ASAT techniques would be to loft a payload of 1-2 cm metal spheres directly from the ground and scatter them into the path of a satellite; the impact of one or a few spheres, at a satellite's orbital velocity of 8 km/sec, would do severe damage to all but an armoured satellite. In a sense, this is the attack mode of the Soviet ASAT interceptor—its high-explosive warhead detonates as the ASAT nears its target, destroying it with a shower of metal fragments. A capability to orbit satellites would not be required for this mode of attack; a rocket on direct ascent, with several hundred kilograms of payload, would seem sufficient.

In fact, if ASAT attacks were quite this simple, it would be difficult to explain why the Soviet interceptor is such a gigantic weapon (roughly 6 metres long, 2500 kg in weight, launched by a booster weighing 180 000 kg), or why the United States is spending almost $4 billion to acquire its ASAT. The actual destruction of a satellite target lies at the end of a chain of events, each of which poses substantial technical challenges. We will consider here the requirements for direct-ascent or orbital ASAT interceptors, because these seem the most likely forms of a third party ASAT. All these tasks would apply equally to space mines, which in effect are long-lived orbiting ASAT interceptors.

To begin with, the attacker must predict the location of the target quite precisely, so that the ASAT can be brought into position. Ground-based radars, lasers or optical telescopes can detect and track satellites, but the attacker needs to identify which of many satellites passing overhead is the target, and none of these devices can provide that degree of detail. The attacker must either keep track of all satellites from launch, get information from someone who does, or find some distinct characteristic to identify the target. To provide the most precise data, a tracking network should have several radars, lasers, or telescopes geographically spaced and triangulated on the target for several orbits; against manoeuvring or newly-launched satellites, even this network will provide inaccurate data. In the best of circumstances, errors in the predicted location of a target will still be tens or hundreds of metres.

The ASAT must also be brought to the target's path with precise timing and accuracy: if the ASAT arrives at the intercept point less than a half-second early or late, it will miss the target by a kilometre or more. A highly reliable booster, well-trained launch team and highly accurate missile guidance are indispensable. Nonetheless, the accumulated errors are likely to be quite large, and in the final moments of high-velocity approach to the target, the ASAT must either guide itself to the target with a homing system or have a warhead with great lethal range. As a practical matter, it is difficult to loft a non-nuclear warhead that would have a lethal radius in all directions of a kilometre or more. The idea of scattering metal spheres in a target's path has great simplicity, but filling a volume of 1 km radius with enough metal spheres to assure damage to a passing target would require a warhead of extraordinary weight. For non-nuclear attacks, highly accurate guidance, homing and warhead aiming are essential.

The ASAT programmes of the two superpowers represent very different approaches to the problems of attacking satellites, and both demonstrate the difficulties. The Soviet system was first designed and tested in the 1960s, so it represents ASAT techniques that might be achievable by a third party at the earliest stages of a space programme. The US ASAT is a product of the 1980s, so it represents newer techniques that more modern technology makes possible. Both superpowers have extensive, global space tracking systems, which can track and catalogue all potential ASAT targets with great precision.

Even so, both ASATs depend upon homing systems to compensate for tracking and booster errors. The Soviet ASAT uses an on-board radar to home on its target. By going into orbit before attacking, the ASAT can reduce the velocity differential with its target and thus gain time for final manoeuvring. And since the radar pin-points the target, presumably the ASAT can aim its warhead squarely at the target, and thus get by with a somewhat smaller fragmentation warhead. Nonetheless, the radar antenna, electronics, batteries and warhead all add mass that must be manoeuvred to the target. Steering motors, fuel tanks, fuel and so on in turn add more mass that must also be manoeuvred to the target; the end result is an ASAT weighing several tonnes. Even though it exploited the best technology available to a superpower at the early stages of its space programme, the success against targets in fairly undemanding tests has been no greater than 50-70 per cent.

The US ASAT homes on its target with a passive infra-red sensor, so it does not need massive electronics, batteries or antenna to send and receive a radar signal. In turn, this means its manoeuvring motors can be very small. The ASAT is launched directly at the target and does not go into orbit, so its closing velocity with the target is very high. This means it can kill the target by direct impact, thus avoiding the weight and complexity of a warhead. Because the ASAT is smaller, its booster can be much smaller. All this 'simplicity' comes at a price, however. At high closing velocities, the ASAT has little time for homing manoeuvres, so its booster guidance and information on target location must be very accurate. Put differently, the ASAT must be supported by very sophisticated missile- and space-tracking technologies. And the ASAT itself is technically complicated; the current cost of each ASAT weapon is about $25 million.

The basic building blocks for the ASAT mission are becoming more widely available to third parties, but are still limited. For instance, the number of non-superpowers that have an independent national capacity to loft payloads to orbit has grown, but remains few: China, France, India, Japan and the UK. It is difficult to predict how many more states may acquire space boosters in the coming 10 years or so. But if past experience is a guide, the most likely states are those that already have advanced research rocket programmes, for example, Canada and Brazil.[19]

Similarly, satellite-tracking capabilities have become more widespread, in some instances fostered by the superpowers themselves. The United States has signed several dozen bilateral pacts with other states to establish space-tracking sites that support satellite and manned space programmes, and NASA has encouraged third party participation in site operations and has trained local personnel. The Soviet Union has from time to time announced the establishment of tracking stations in various East European, Caribbean, African and Middle Eastern states. In 1979 and 1981, the Soviets granted access to a tracking site near Moscow for use by Indian experts on occasions when the Soviets were launching the Indian Bhaskara satellites.[20]

As third party space programmes have become more ambitious, so has interest in widely-scattered tracking facilities. Japan maintains tracking facilities at its Tanegashima Island launch site, at Ogasawara north of Iwo Jima, at Katsura near Tokyo, and on Okinawa. Italy has a tracking site at Malindi, Kenya. To support Ariane, the ESA originally established tracking stations at its Kourou launch site in Guiana and at Akakro, Ivory Coast. To accommodate advanced versions of Ariane with longer booster burn times, additional tracking stations have become necessary, leading ESA to study Gabon, Congo Brazzaville, and Italy's Malindi facility as future tracking sites. Third party launches to polar orbits will demand added tracking services, and Sweden has found China, France, India and Japan all interested in using its space track site at Esrange, above the Arctic Circle. And just as superpower space programmes have helped spread tracking technologies, third parties may also aid other third parties in acquiring a space track capacity. ESA, for instance, constructed a tracking facility in Natal, Brazil, which it then turned over to Brazilian control in exchange for continued tracking services for Ariane.[21]

A state able to track its own satellites may not be able to track potential ASAT targets, however. Generally, satellites are equipped by the launching state with devices that aid tracking (special transponders, distinctive radio signals or beacons, laser reflectors, etc.), and ground 'tracking' systems may only be telemetry stations that are dependent upon a satellite's emitted signals to detect and track. Military satellites of other states—the most likely ASAT targets—may not be so co-operative.

Finally, any third party that attempts ASAT attacks on the satellites of a superpower must anticipate retaliation against its own satellites. In a world in which the superpowers are testing and deploying advanced BMD, even imperfect defences will be highly effective ASATs. Hence, the superpowers may well be equipped to destroy third party satellites, in large numbers and at great altitudes. Moreover, the superpowers are likely to have greater capacity to replace their own disrupted satellite services promptly, either with newly-launched satellite replacements or ground-based alternative systems.

To what do all these factors add up? Very few states have acquired the technological building blocks for an ASAT weapon. The space programmes of a few third states may become quite sophisticated before the end of the century, encompassing reusable space shuttles, manned space laboratories, and independent, manoeuvring space vehicles. None of these programmes, designed for commercial or scientific purposes, will yield a highly reliable, militarily effective ASAT directly, however. At best they might provide equipment that could be jury-rigged into a lesser, 'nuisance' ASAT (e.g., a system unable to threaten more than one or two low-orbit satellites). Dire predictions that widespread ASAT threats will emerge from ordinary space programmes have little basis in fact. The general trend in civilian space programmes is toward large, high-cost spacecraft, optimally designed for their specific purposes, acting in co-operation with other 'friendly' spacecraft—precisely the opposite of what the ASAT task requires. If this disappoints those who hoped such predictions would compel the superpowers to behave themselves in outer space, it ought to encourage those who still hope for ASAT arms control. Acquiring an effective ASAT weapon demands a deliberate, political decision.

It would be regrettable if the superpowers were to underestimate the influence they still have over the decisions of third parties. At the moment, acquiring an ASAT would be incompatible with the professed principles and political associations of all but perhaps one or two of the ASAT-potential states. The high cost of an ASAT programme is an added inhibition. By continuing to offer open and generous access to their own space services, the superpowers can reduce the incentive of third states to go off on their own and acquire independent capacities that could be turned to ASAT purposes. And by expanding co-operative ventures among states on space programmes, the superpowers can cultivate the political connections that discourage independent national ASAT programmes.

The price for gaining this restraint in third party ASAT programmes is that the superpowers must prove they are responsive to the interests of third parties in outer space and are willing to restrain their own programmes when they infringe on those interests. It would do the superpowers little harm to pay such a modest price.

Notes and references

1 Donald Hafner is an associate professor at Boston College. Prof. Hafner served as an adviser with the US SALT II delegation; as executive secretary of Guidance Committees for the SALT and ASAT arms control delegations; and as an ACDA analyst on the National Security Council SALT and ASAT working groups.

2 For a review of superpower attitudes, see O'Neill, P., 'The development of the international law of outer space', and Russell, M., 'The Soviet legal view of military space activities', in W. Durch, ed., *National Interests and Military Use of Space* (Ballinger: Cambridge, MA, 1984).

3 For a detailed account of US and Soviet ASAT programmes, see Stares, P., *Space and National Security* (The Brookings Institution: Washington, DC, 1987, forthcoming).

4 Velikhov, Y.Y., Sagdeev, R. and Kokoshin, A., eds, *Weaponry in Space: The Dilemma of Security* (Mir Publishers: Moscow, 1986), pp. 99-100.

5 Each superpower will attempt to harden its own satellites against attack, and knowing the other is doing the same, each will seek to develop ASAT weapons with ever greater lethal effect that can overcome hardening. Since third parties may lack the techniques or resources for hardening their own satellites, these may be vulnerable to superpower ASAT effects, even at long distances from an actual ASAT test. For example, by exploiting known techniques, the superpowers could harden the electronics of their satellites to withstand a very large nuclear ASAT detonated less than 10 km away, while an unhardened satellite would be disabled at distances out to several hundred kilometres from the detonation.

6 US Congress, Office of Technology Assessment, *Anti-Satellite Weapons, Countermeasures, and Arms Control* (US Government Printing Office: Washington, DC, Sep. 1985), pp. 49-62; US Congress, Congressional Research Service, *Space Activities of the United States, Soviet Union, and Other Launching Countries* (US Government Printing Office: Washington, DC, Feb. 1982), pp. 19-20.

7 Smaller pieces of debris from ASAT tests (less than 0.1 cm), too small to be tracked by radar but potentially damaging to spacecraft, can be estimated at perhaps 40 000. I am grateful to Prof. George J. Flynn, of the State University of New York at Plattsburgh, for bringing these estimates to my attention.

8 The test region could extend both above and below the orbiting BMD weapons. ICBM warheads reach altitudes of 1000 km on normal trajectories, yet the orbital altitude for space-based BMD weapons is likely to be lower. Hence space-based BMD weapons firing at warheads in mid-course may well be firing 'upward', posing some hazard to satellites at higher altitudes. The same would be true, of course, for ground-based laser BMD weapons.

9   For a discussion of the legal priciples involved in closing off ocean areas during tests, see O'Connell, D. P., *The International Law of the Sea* (Clarendon Press: Oxford, Manchester, 1984), vol. II, pp. 809-13; and Churchill, R.R. and Lowe, A.V., *The Law of the Sea* (Manchester University Press: Manchester, 1983), pp. 147-8.

10   In July 1983, a window of the US space shuttle Challenger was struck by a piece of debris and fractured badly enough to require replacement. Subsequent analysis revealed that the piece of debris was a chip of paint, estimated at no more than 0.2 mm in diameter. See 'Space junk grows with weapons tests', *Science,* 25 Oct. 1985.

11   It has been estimated, for instance, that an object the size of the US space shuttle faces only one chance in a million, per mission, of being hit by a large piece of debris (larger than 10 cm). On the other hand, there are at least four instances in recent years when satellites suffered catastrophic damage from apparent collisions with debris (the US PAGEOS in 1975; the European GEOS-2 in 1978; and the Soviet Kosmos 954 in 1977 and Kosmos 1275 in 1981). See 'Space junk grows with weapons tests' (note 10).

12   'Space defense operations center upgrades assessment capabilities', *Aviation Week & Space Technology*, 9 Dec. 1985, p. 73; and Marshall, E., 'Space junk grows with weapons tests' (note 10).

13   For a more detailed discussion of arms control for ASAT weapons, see Hafner, D., 'Potential negotiated measures for ASAT arms control', in the Report of the Aspen Strategy Group, *Seeking Stability in Space* (University Press of America: Lanham, Maryland, forthcoming); and Hafner, D., 'Anti-Satellite Weapons: the prospects for arms control', in B. Jasani, ed., *Outer Space: A New Dimension of the Arms Race*, SIPRI (Taylor & Francis: London, 1982).

14   Under the 1974 Convention on Registration of Objects Launched Into Outer Space, states are supposed to report all objects they place in space, along with the object's initial orbital parameters and general function. Large pieces of debris (e.g., payload shrouds and spent boosters) are registered by the superpowers, but over time the orbital parameters of the object can change. Complete cataloguing and computation of current orbital parameters for several thousand registered space objects is an enormous undertaking, and few states could afford the investment in facilities that are required.

15   An exception would be BMD survival techniques that rely upon 'stealth' satellites, left inert and concealed during peace-time and activated only when needed in crisis. Manoeuvring these to avoid collisions might reveal their position and purpose, so protecting them might require 'hiding' them near other, visible BMD satellites and establishing keep-out zones from which all parties would be excluded.

16   During the past several years, Soviet legal specialists have asserted a state's right of self-defence if close approaches by the space objects of other states would damage or interfere with its own spacecraft, and have discussed the concept of 'security zones' around spacecraft that could be entered by third parties

only with explicit permission. See Russell, M., 'The Soviet legal view of military space activities', in W. Durch, ed., *National Interests and the Military Use of Space* (Ballinger: Cambridge, Mass., 1984), pp. 215-16. In the US, a fairly elaborate proposal for 'self-defense zones' as a safeguard against ASATs has been made by Albert Wohlstetter and Brian Chow. In their formulation, the geosynchronous orbital band would be divided into perhaps 36 zones, with 12 for NATO, 12 for the Warsaw Pact and 12 for all neutral nations. No more than two satellites could be stationed by one of these groups in the zone of another; transits of spacecraft through a zone would be regulated. Any object violating these rules would be subject to instant destruction. See 'Arms control that could work', *Wall Street Journal,* 17 July 1985.

17 The Soviet Academy of Sciences report on space-based BMD, for instance, argues that the very act of protecting itself, even against inert debris, might require bringing a BMD battle station up to 'wartime' status. In turn, this might provoke the other side's sytem into wartime status as well, thus generating a very dangerous cycle. See Y. Velikhov, R. Sagdeev, and A. Kokoshin, eds., *Weaponry in Space: The Dilemma of Security* (Mir Publishers: Moscow, 1986) pp. 61-2.

18 See, for instance, President Reagan's *Report to the Congress on US Policy on ASAT Arms Control,* 31 March 1984: 'For example, any nation routinely conducting space rendezvous and docking maneuvers...could, under the guise of that activity, develop spacecraft equipped to maneuver into the path of, or detonate next to, another nation's spacecraft'. '...in general, many activities related to space give rise to capabilities inherently useful for ASAT purposes...'. See also 'Report supports exchange of space information with Soviets if done cautiously', *The Washington Times,* 18 July 1985: 'The technology used in the Soyuz and Apollo spacecrafts in 1975 is similar to that used in anti-satellite warfare, Mr Codevilla said [Angelo Codevilla, aide to Senator Malcolm Wallop of Wyoming]. The relationship between civil and military space applications is very close and therefore space research "has important military consequences", Assistant Secretary of Defense Richard Perle said. "One of the reasons why it is so difficult to come up with arms control limitations on space activities is that it is virtually impossible to determine the intent just from observing the activity", Mr. Perle said'.

19 Canada's *Black Brant* research rocket can carry 275 kg to 390 km or 70 kg to 1500 km; Brazil's *Sonda 2/B* can loft 65 kg to 180 km; Poland's *Meteor 2K* can loft 10 kg to 100 km (see 'International research rockets', *Aviation Week & Space Technology,* 10 Mar. 1986, p. 177). In 1984 Brazil announced its intention to develop a space booster able to orbit satellites by 1990 (see 'Brazil plans to launch own satellites', *Aviation Week & Space Technology,* 9 July 1984, pp. 60-1). Thirteen members of the ESA in principle have Ariane available as an ASAT booster. Boosters might become available for purchase from commercial firms in the future, but it seems unlikely that a state seeking an independent ASAT capability would rely upon a foreign commercial firm for its boosters.

20   US Congress, Office of Technology Assessment, *International Cooperation and Competition in Civilian Space Activities* (US Government Printing Office: Washington, DC, 1985), pp. 35-41; US Congress, Senate, *Soviet Space Programs: 1976-80* (US Government Printing Office: Washington, DC, 1982), Part 1, pp. 121-5.

21   See 'ESA seeks US tracking agreement, new downrange station for Ariane', *Aviation Week & Space Technology*, 7 Jan. 1985, p. 117; 'Swedes study new role for Esrange base', *Aviation Week & Space Technology*, 15 Apr. 1985, p. 55; 'Tsukuba expanding mission control, vehicle checkout facilities', *Aviation Week & Space Technology*', 28 July 1986, p. 40. In some measure, the superpowers may be able to slow the diffusion of space tracking capabilities by offering their own services on a co-operative basis. The US Space Defense Command Center (SPADOC), which tracks and catalogues all space objects, provides information and warnings on potential collisions between space launches and orbiting objects to other co-operating states with space programmes, thus reducing those states' requirements for independent tracking data. And the ready availability of NASA's extensive tracking network has probably reduced the incentive for third parties to establish their own independent sites. Japan, for example, exploits NASA facilities at Christmas Island and Santiago, Chile, and ESA has relied on NASA's Ascension Island site and will use NASA's Wallops Island and Bermuda facilities for polar launches. But NASA intends to rely increasingly on its Tracking and Data Relay Satellites (TDS) as an alternative to ground stations for communication and control of its space programmes. As NASA's ground facilities are closed down, third parties may build more of their own—or acquire those NASA is abandoning.

Appendix

Treaty between the United States of America and the Union of Soviet Socialist Republics on the limitation of anti-ballistic missile systems (ABM Treaty, 1972)

SIGNED: Moscow, 26 May 1972
ENTERED INTO FORCE: 3 October 1972

The United States of America and the Union of Soviet Socialist Republics, hereinafter referred to as the Parties,

Proceeding from the premise that nuclear war would have devastating consequences for all mankind,

Considering that effective measures to limit anti-ballistic missile systems would be a substantial factor in curbing the race in strategic offensive arms and would lead to a decrease in the risk of outbreak of war involving nuclear weapons,

Proceeding from the premise that the limitation of anti-ballistic missile systems, as well as certain agreed measures with respect to the limitation of strategic offensive arms, would contribute to the creation of more favorable conditions for further negotiations on limiting strategic arms,

Mindful of their obligations under Article VI of the Treaty on the Non-Proliferation of Nuclear Weapons,

Declaring their intention to achieve at the earliest possible date the cessation of the nuclear arms race and to take effective measures towards reductions in strategic arms, nuclear disarmament, and general and complete disarmament,

Desiring to contribute to the relaxation of international tension and the strengthening of trust between States,

Have agreed as follows:

Article I

1. Each party undertakes to limit anti-ballistic missile (ABM) systems and to adopt other measures in accordance with the provisions of this Treaty.

2. Each Party undertakes not to deploy ABM systems for a defense of the territory of its country and not to provide a base for such a defense, and not to deploy ABM systems for defense of an individual region except as provided for in Article III of this Treaty.

Article II

1. For the purpose of this Treaty an ABM system is a system to counter strategic ballistic missiles of their elements in flight trajectory, currently consisting of:

(*a*) ABM interceptor missiles, which are interceptor missiles constructed and deployed for an ABM role, or of a type tested in an ABM mode;

(*b*) ABM launchers, which are launchers constructed and deployed for launching ABM interceptor missiles; and

(*c*) ABM radars, which are radars constructed and deployed for an ABM role, or of a type tested in an ABM mode.

2. The ABM system components listed in paragraph 1 of this Article include those which are:

(*a*) operational;

(*b*) under construction;

(*c*) undergoing testing;

(*d*) undergoing overhaul, repair or conversion; or

(*e*) mothballed.

Article III

Each Party undertakes not to deploy ABM systems or their components except that:

(*a*) within one ABM system deployment area having a radius of one hundred and fifty kilometers and centered on the Party's national capital, a Party may deploy: (1) no more than one hundred ABM launchers and no more than one hundred ABM interceptor missiles at launch sites, and (2) ABM radars within no more than six ABM radar complexes, the area of each complex being circular and having a diameter of no more than three kilometers; and

(*b*) within one ABM system deployment area having a radius of one hundred and fifty kilometers and containing ICBM silo launchers, a Party may deploy: (1) no more than one hundred ABM launchers and no more than one hundred ABM radars comparable in potential to corresponding ABM radars operational or under construction on the date of signature of the Treaty in an ABM system deployment area containing ICBM silo launchers, and (3) no more than eighteen ABM radars each having a potential less than the potential of the smaller of the above-mentioned two large phased-array ABM radars.

Article IV

The limitations provided for in Article III shall not apply to ABM systems or their components used for development or testing, and located within current or additionally agreed test ranges. Each Party may have no more than a total of fifteen ABM launchers at test ranges.

Article V

1. Each Party undertakes not to develop, test, or deploy ABM systems or components which are sea-based, air-based, space-based or mobile land-based.

2. Each Party undertakes not to develop, test, or deploy ABM launchers for launching more than one ABM interceptor missile at a time from each launcher, not to modify deployed launchers to provide them with such a capability, not to develop, test, or deploy automatic or semi-automatic or other similar systems for rapid reload of ABM launchers.

Article VI

To enhance assurance of the effectiveness of the limitations on ABM systems and their components provided by the Treaty, each Party undertakes:

(*a*) not to give missiles, launchers, or radars, other than ABM interceptor missiles, ABM launchers, or ABM radars, capabilities to counter strategic ballistic missiles or their elements in flight trajectory, and not to test them in an ABM mode; and

(*b*) not to deploy in the future radars for early warning of strategic ballistic missile attack except at locations along the periphery of its national territory and orientated outward.

Article VII

Subject to the provisions of this Treaty, modernization and replacement of ABM systems or their components may be carried out.

Article VIII

ABM systems or their components in excess of the numbers or outside the areas specified in this Treaty, as well as ABM systems or their components prohibited by this Treaty, shall be destroyed or dismantled under agreed procedures within the shortest possible agreed period of time.

Article IX

To assure the viability and effectiveness of this Treaty, each Party undertakes not to transfer to other States, and not to deploy outside its national territory, ABM systems or their components limited by this Treaty.

Article X

Each Party undertakes not to assume any international obligations which would conflict with this Treaty.

Article XI

The Parties undertake to continue active negotiations for limitations on strategic offensive arms.

Article XII

1. For the purpose of providing assurance of compliance with the provisions of this Treaty, each Party shall use national technical means of verification as its disposal in a manner consistent with generally recognized principles of international law.

2. Each Party undertakes not to interfere with the national technical means of verification of the other Party operating in accordance with paragraph 1 of this Article.

3. Each Party undertakes not to use deliberate concealment measures which impede verification by national technical means of compliance with the provisions of this Treaty. This obligation shall not require changes in current construction, assembly, conversion, or overhaul practices.

Article XIII

1. To promote the objectives and implementation of the provisions of this Treaty, the Parties shall establish promptly a Standing Consultative Commission, within the framework of which they will:

(*a*) consider questions concerning compliance with the obligations assumed and related situations which may be considered ambiguous;

(*b*) provide on a voluntary basis such information as either Party considers necessary to assure confidence in compliance with the obligations assumed;

(*c*) consider questions involving unintended interference with national technical means of verification;

(*d*) consider possible changes in the strategic situation which have a bearing on the provisions of this Treaty;

(*e*) agree upon procedures and dates for destruction or dismantling of ABM systems o their components in cases provided for by the provisions of this Treaty;

(*f*) consider, as appropriate, possible proposals for further increasing the viability of this Treaty; including proposals for amendments in accordance with the provisions of this Treaty;

(*g*) consider, as appropriate, proposals for further measures aimed at limiting strategic arms.

2. The Parties through consultation shall establish, and may amend as appropriate, Regulations for the Standing Consultative Commission governing procedures, composition and other relevant matters.

Article XIV

1. Each Party may propose amendments to this Treaty. Agreed amendments shall enter into force in accordance with the procedures governing the entry into force of this Treaty.

2. Five years after entry into force of this Treaty, and at five-year intervals thereafter, the Parties shall together conduct a review of this Treaty.

Article XV

1. This Treaty shall be of unlimited duration.

2. Each Party shall, in exercising its national sovereignty, have the right to withdraw from this Treaty if it decides that extraordinary events related to the subject matter of this Treaty have jeopardized its supreme interests. It shall give notice of its decision to the other Party six months prior to withdrawal from the Treaty. Such notice shall include a statement of the extraordinary events the notifying Party regards as having jeopardized its supreme interests.

Article XVI

1. This Treaty shall be subject to ratification in accordance with the constitutional procedures of each Party. The Treaty shall enter into force on the day of the exchange of instruments of ratification.

2. This Treaty shall be registered pursuant to Article 102 of the Charter of the United Nations.

DONE at Moscow on May 26, 1972, in two copies, each in the English and Russian languages, both texts being equally authentic.

For the United States of America

President of the United States of America

For the Union of Soviet Socialist Republics

General Secretary of the Central Committee of the CPSU

Agreed statements, common understandings, and unilateral statements regarding the treaty between the United States of America and the Union of Soviet Socialist Republics on the limitation of anti-ballistic missile systems

Agreed statements

The document set forth below was agreed upon and initialed by the Heads of the Delegations on May 26, 1972 [letter designations added]:

[A]
The Parties understand that, in addition to the ABM radars which may be deployed in accordance with subparagraph (a) of Article III of the Treaty, those non-phased-array ABM radars operational on the date of signature of the Treaty within the ABM system deployment area for defense of the national capital may be retained.

[B]
The Parties understand that the potential (the product of mean emitted power in watts and antenna area in square meters) of the smaller of the two large phased-array ABM radars referred to in subparagraph (b) of Article III of the Treaty is considered for purposes of the Treaty to be three million.

[C]
The Parties understand that the centre of the ABM system deployment area centered on the national capital and the centre of the ABM system deployment area containing ICBM silo launchers for each Party shall be separated by no less than thirteen hundred kilometers.

[D]
In order to insure fulfillment of the obligation not to deploy ABM systems and their components except as provided in Article III of the Treaty, the Parties agree that in the event ABM systems based on other physical principles and including components capable of substituting for ABM interceptor missiles, ABM launchers,

or ABM radars are created in the future, specific limitations on such systems and their components would be subject to discussion in accordance with Article XIII and agreement in accordance with Article XIV of the Treaty.

[E]
The Parties understand that Article V of the Treaty includes obligations not to develop, test or deploy ABM interceptor missiles for the delivery by each ABM interceptor missile of more than one independently guided warhead.

[F]
The Parties agree not to deploy phased-array radars having a potential (the product of mean emitted power in watts and antenna area in square meters) exceeding three million, except as provided for in Articles III, IV, and VI of the Treaty, or except for the purposes of tracking objects in outer space or for use as national technical means of verification.

[G]
The Parties understand that Article IX of the Treaty includes the obligation of the US and the USSR not to provide to other States technical descriptions or blue prints specially worked out for the construction of ABM systems and their components limited by the Treaty.

Common understandings

Common understanding of the Parties on the following matters was reached during the negotiations:

A. Location of ICBM defenses

The U.S. Delegation ma₋₋ the following statement on May 26, 1972:

Article III of the ABM Treaty provides for each side one ABM system deployment area centered on its national capital and one ABM system deployment area containing ICBM silo launchers. The two sides have registered agreement on the following statement: 'The Parties understand that the center of the ABM system deployment area centered on the national capital and the center of the ABM system deployment area containing ICBM silo launchers for each Party shall be separated by no less than thirteen hundred kilometers.' In this connection, the U.S. side notes that its ABM system deployment area for defense of ICBM silo launchers, located west of the Mississippi River, will be centered in the Grand Forks

ICBM silo launcher deployment area. (See Agreed Statement [C].)

B. ABM test ranges

The U.S. Delegation made the following statement on April 26, 1972:

Article IV of the ABM Treaty provides that 'the limitations provided for in Article III shall not apply to ABM systems or their components used for development or testing, and located within current or additionally agreed test ranges.' We believe it would be useful to assure that there is no misunderstanding as to current ABM test ranges. It is our understanding that ABM test ranges encompass the area within which ABM components are located for test purposes. The current U.S. ABM test ranges are at White Sands, New Mexico, and at Kwajalein Atoll, and the current Soviet ABM test range is near Sary Shagan in Kazakhstan. We consider that non-phased array radars of types used for range safety or instrumentation purposes may be located outside of ABM test ranges. We interpret the reference in Article IV to 'additionally agreed test ranges' to mean that ABM components will not be located at any other test ranges without prior agreement between our Governments that there will be such additional ABM test ranges.

On May 5, 1972, the Soviet Delegation stated that there was a common understanding on what ABM test ranges were, that the use of the types of non-ABM radars for range safety or instrumentation was not limited under the Treaty, that the reference in Article IV to 'additionally agreed' test ranges was sufficiently clear, and that national means permitted identifying current test ranges.

C. Mobile ABM sytems

On January 29, 1972, the U.S. Delegation made the following statement:

Article V(1) of the Joint Draft Text of the ABM Treaty includes an undertaking not to develop, test, or deploy mobile land-based ABM systems and their components. On May 5, 1971, the U.S. side indicated that, in its view, a prohibition on deployment of mobile ABM systems and components would rule out the deployment of ABM launchers and radars which were not permanent fixed types. At the time, we asked for the Soviet view of this interpretation. Does the Soviet side agree with the U.S. side's interpretation put forward on May 5, 1971?

On April 13, 1972, the Soviet Delegation said there is a general common understanding on this matter.

D. Standing Consultative Commission

Ambassador Smith made the following statement on May 22, 1972:

The United States proposes that the sides agree that, with regard to initial implementation of the ABM Treaty's Article XIII on the Standing Consultative Commission (SCC) and of the consultation Articles to the Interim Agreement on offensive arms and the Accidents Agreement,[1] agreement establishing the SCC will be worked out early in the follow-on SALT negotiations; until that is completed, the following arrangements will prevail: when SALT is in session, any consultation desired by either side under these Articles can be carried out by the two SALT Delegations; when SALT is not in session, *ad hoc* arrangements for any desired consultations under these Articles may be made through diplomatic channels.

Minister Semenov replied that, on an *ad referendum* basis, he could agree that the U.S. statement corresponded to the Soviet understanding.

E. Standstill

On May 6, 1972, Minister Semenov made the following statement:

In an effort to accommodate the wishes of the U.S. side, the Soviet Delegation is prepared to proceed on the basis that the two sides will in fact observe the obligations of both the Interim Agreement and the ABM Treaty beginning from the date of signature of these two documents.

In reply, the U.S. Delegation made the following statement on May 20, 1972:

The U.S. agrees in principle with the Soviet statement made on May 6 concerning observance of obligations beginning from date of signature but we would like to make clear our understanding that this means that, pending ratification and acceptance, neither

side would take any action prohibited by the agreements after they had entered into force. This understanding would continue to apply in the absence of notification by either signatory of its intention not to proceed with ratification or approval.

The Soviet Delegation indicated agreement with the U.S. statement.

Unilateral statements

The following noteworthy unilateral statements were made during the negotiations by the United States Delegation:

A. Withdrawal from the ABM Treaty

On May 9, 1972, Ambassador Smith made the following statement:

The U.S. Delegation has stressed the importance the U.S. Government attaches to achieving agreement on more complete limitations on strategic offensive arms, following agreement on an ABM Treaty and on an Interim Agreement on certain measures with respect to the limitation of strategic offensive arms. The U.S. Delegation believes that an objective of the follow-on negotiations should be to constrain and reduce on a long-term basis threats to the survivability of our respective strategic retaliatory forces. The USSR Delegation has also indicated that the objectives of SALT would remain unfulfilled without the achievement of an agreement providing for more complete limitations on strategic offensive arms. Both sides recognize that the initial agreements would be steps towards the achievement of more complete limitations on strategic arms. If an agreement providing for more complete strategic offensive arms limitations were not achieved within five years, U.S. supreme interests could be jeopardized. Should that occur, it would constitute a basis for withdrawal from the ABM Treaty. The U.S. does not wish to see such a situation occur, nor do we believe that the USSR does. It is because we wish to prevent such a situation that we emphasize the importance the U.S. Government attaches to achievement of more complete limitations on strategic offensive arms. The U.S. Executive will inform the Congress, in connection with Congressional consideration of the ABM Treaty and the Interim

---

[1] See Article 7 of Agreement to Reduce the Risk of Outbreak of Nuclear War Between the United States of America and the Union of Soviet Socialist Republics, signed 30 September, 1971.

Agreement, of this statement of the U.S. position.

B. Tested in ABM mode

On April 7, 1972, the U.S. Delegation made the following statement:

Article II of the Joint Text Draft uses the term 'tested in an ABM mode,' in defining ABM components, and Article VI includes certain obligations concerning such testing. We believe that the sides should have a common understanding of this phrase. First, we would note that the testing provisions of the ABM Treaty are intended to apply to testing which occurs after the date of signature of the Treaty, and not to any testing which may have occurred in the past. Next, we would amplify the remarks we have made on this subject during the previous Helsinki phase by setting forth the objectives which govern the U.S. view on the subject, namely, while prohibiting testing of non-ABM components for ABM purposes: not to prevent testing of ABM components, and not to prevent testing of non-ABM components for non-ABM purposes. To clarify our interpretation of 'tested in an ABM mode,' we note that we would consider a launcher, missile or radar to be 'tested in an ABM mode' if, for example, any of the following events occur: (1) a launcher is used to launch an ABM interceptor missile, (2) an interceptor missile is flight tested against a target vehicle which has a flight trajectory with characteristics of a strategic ballistic missile flight trajectory, or is flight tested in conjunction with the test of an ABM interceptor missile or an ABM radar at the same test range, or is flight tested to an altitude inconsistent with interception of targets against which air defenses are deployed, (3) a radar makes measurements on a cooperative target vehicle of the kind referred to in item (2) above during the reentry portion of its trajectory or makes measurements in conjunction with the test of an ABM interceptor missile or an ABM radar at the same test range. Radars used for purposes such as range safety or instrumentation would be exempt from application of these criteria.

C. No-transfer article of ABM Treaty

On April 18, 1972, the U.S. Delegation made the following statement:

In regard to this Article [IX], I have a brief

and I believe self-explanatory statement to make. The U.S. side wishes to make clear that the provisions of this Article do not set a precedent for whatever provision may be considered for a Treaty on Limiting Strategic Offensive Arms. The question of transfer of strategic offensive arms is a far more complex issue, which may require a different solution.

D. No increase in defense of early warning radars

On July 28, 1970, the U.S. Delegation made the following statement:

Since Hen House radars [Soviet ballistic missile early warning radars] can detect and track ballistic missile warheads at great distances, they have a significant ABM potential. Accordingly, the U.S. would regard any increase in the defenses of such radars by surface-to-air missiles as inconsistent with an agreement.

Protocol to the Treaty Between the United States of America and the Union of Soviet Socialist Republics on the Limitation of Anti-Ballistic Missile Systems, July 3, 1974 (ABM Treaty Protocol, 1974)

The United States of America and the Union of Soviet Socialist Republics, hereinafter referred to as the Parties,

Proceeding from the Basic Principles of Relations between the United States of America and the Union of Soviet Socialist Republics signed on May 29, 1972,

Desiring to further the objectives of the Treaty between the United States of America and the Union of Soviet Socialist Republics on the Limitation of Anti-Ballistic Missile Systems signed on May 26, 1972, hereinafter referred to as the Treaty,

Reaffirming their conviction that the adoption of further measures for the limitation of strategic arms would contribute to strengthening international peace and security,

Proceeding from the premise that further limitation of anti-ballistic missile systems will create more favorable conditions for the completion of work on a permanent agreement on more complete measures for the limitation of strategic offensive arms,

Have agreed as follows:

Article I

1. Each Party shall be limited at any one time to a single area out of the two provided in Article III of the Treaty for deployment of anti-ballistic missile (ABM) systems or their components and accordingly shall not exercise its right to deploy an ABM system or its components in the second of the two ABM system deployment areas permitted by Article III of the Treaty, except as an exchange of one permitted area for the other in accordance with Article II of this Protocol.

2. Accordingly, except as permitted by Article II of this Protocol: the United States of America shall not deploy an ABM system or its components in the area centered on its capital, as permitted by Article III(a) of the Treaty, and the Soviet Union shall not deploy an ABM system or its components in the deployment area of intercontinental ballistic missile (ICBM) silo launchers permitted by Article III(b) of the Treaty.

Article II

1. Each Party shall have the right to dismantle or destroy its ABM system and the components thereof in the area where they are presently deployed and to deploy an ABM system or its components in the alternative area permitted by Article III of the Treaty, provided that prior to initiation of construction, notification is given in accord with the procedure agreed to by the Standing Consultative Commission, during the year beginning October 3, 1977, and ending October 2, 1978, or during any year which commences at five year intervals thereafter, those being the years for periodic review of the Treaty, as provided in Article XIV of the Treaty. This right may be exercised only once.

2. Accordingly, in the event of such notice, the United States would have the right to dismantle or destroy the ABM system and its components in the deployment area of ICBM silo launchers and to deploy an ABM system or its components in an area centered on its capital, as permitted by Article III(a) of the Treaty, and the Soviet Union would have the right to dismantle or destroy the ABM system and its components in the area centered on its capital and to deploy an ABM system or its components in an area containing ICBM silo launchers, as permitted by Article III(b) of the Treaty.

3. Dismanting or destruction and deployment of ABM systems or their components and the notification thereof shall be carried out in accordance with Article VIII of the ABM Treaty and procedures agreed to in the Standing Consultative Commission.

Article III

The rights and obligations established by Treaty remain in force and shall be complied with by the Parties except to the extent modified by this Protocol. In particular, the deployment of an ABM system or its components within the area selected shall remain limited by the levels and other requirements established by the Treaty.

Article IV

This Protocol shall be subject to ratification in accordance with the constitutional procedures of each Party. It shall enter into force on the day of the exchange of instruments of ratification and shall thereafter be considered an integral part of the Treaty.

Done at Moscow on July 3, 1974, in duplicate, in the English and Russian languages, both texts being equally authentic.

For the United States of America:

RICHARD NIXON
President of the United States of America

For the Union of Soviet Socialist Republics:

L. I. BREZHNEV
General Secretary of the Central Committee of the CPSU

Index